AF753670

Chipless RFID Reader Design
for Ultra-Wideband Technology

Remote Identification Beyond RFID Set

coordinated by
Etienne Perret

# Chipless RFID Reader Design for Ultra-Wideband Technology

*Design, Realization and Characterization*

Marco Garbati
Etienne Perret
Romain Siragusa

ISTE PRESS

ELSEVIER

First published 2018 in Great Britain and the United States by ISTE Press Ltd and Elsevier Ltd

Apart from any fair dealing for the purposes of research or private study, or criticism or review, as permitted under the Copyright, Designs and Patents Act 1988, this publication may only be reproduced, stored or transmitted, in any form or by any means, with the prior permission in writing of the publishers, or in the case of reprographic reproduction in accordance with the terms and licenses issued by the CLA. Enquiries concerning reproduction outside these terms should be sent to the publishers at the undermentioned address:

ISTE Press Ltd
27-37 St George's Road
London SW19 4EU
UK

www.iste.co.uk

Elsevier Ltd
The Boulevard, Langford Lane
Kidlington, Oxford, OX5 1GB
UK

www.elsevier.com

**Notices**
Knowledge and best practice in this field are constantly changing. As new research and experience broaden our understanding, changes in research methods, professional practices, or medical treatment may become necessary.

Practitioners and researchers must always rely on their own experience and knowledge in evaluating and using any information, methods, compounds, or experiments described herein. In using such information or methods they should be mindful of their own safety and the safety of others, including parties for whom they have a professional responsibility.

To the fullest extent of the law, neither the Publisher nor the authors, contributors, or editors, assume any liability for any injury and/or damage to persons or property as a matter of products liability, negligence or otherwise, or from any use or operation of any methods, products, instructions, or ideas contained in the material herein.

MATLAB® is a trademark of The MathWorks, Inc. and is used with permission. The MathWorks does not warrant the accuracy of the text or exercises in this book. This book's use or discussion of MATLAB® software or related products does not constitute endorsement or sponsorship by The MathWorks of a particular pedagogical approach or particular use of the MATLAB® software.

For information on all our publications visit our website at http://store.elsevier.com/

© ISTE Press Ltd 2018
The rights of Marco Garbati, Etienne Perret and Romain Siragusa to be identified as the authors of this work have been asserted by them in accordance with the Copyright, Designs and Patents Act 1988.

British Library Cataloguing-in-Publication Data
A CIP record for this book is available from the British Library
Library of Congress Cataloging in Publication Data
A catalog record for this book is available from the Library of Congress
ISBN 978-1-78548-292-2

# Contents

# Preface

Over the last decade, the advent and diffusion of the smartphone has revolutionized the life of billions of people. The smartphone has the dual advantage of being both a personal computer (PC) and a mobile phone. The mobile phone offers us the opportunity to connect with computers and users, thus exploiting the public telephone network using a low-power portable device. The PC provides us with high computational capability due to its dedicated hardware combined with software suites. Today, these devices are evolving toward a major connectivity in order to meet the incessantly growing request for the exchange of information. The increase of mobility in the era of globalization and the continuous development of technology are changing our life rapidly. Home automation, autonomous transportation, robotics and technology require a strong exchange of information whereby people and machines are connected together in a global network which is commonly known as the Internet of Things (IoT).

The IoT uses different technologies that are involved in personal area network (PAN) wireless communications, such as Bluetooth, WiFi, radio-frequency identification (RFID), infrared, ultra-wideband (UWB) and bar-code reading capability. This book focuses on RFID, which is a technique that allows a reader to receive information from a passive transceiver using the radio-frequency (RF) spectrum. The huge potentiality of this technology lies in its capability of using a totally passive tag (transponder).

The massive use of RFID started in the 1990s and today diverse types of RFID tag operate at different RF bandwidths. All of these devices are in accordance with industrial standards as well as with national and international regulations. RFID tags can incorporate intelligence, execute handshaking protocols, elaborate information and be interconnected with sensors. Consequently, the RFID tags can be used as part of wireless sensor networks (WSNs). The RFID technology is mainly used for the identification of goods by tracking and tracing, in smart logistics and for access control. Recently, it has been incorporated into credit cards.

However, the commonly used identification technology is still the barcode, where the labels are directly printed on objects or paper substrates and read using an optical technique. The barcode was invented in 1948 by Norman Joseph Woodland and Bernard Silver of the University of Drexel and its diffusion started only in 1974 due to improvements in optical technology. Barcode labels can be either unidimensional or bidimensional with many different symbologies. Chipless RFID technology plays an intermediate role between the barcode and classical RFID. It is based on the interrogation of a label (tag) with UWB signals and the subsequent decoding of tag identification (ID) is based on its radar cross section (RCS) signature. The tags are totally passive and do not have any electronic device attached on their surface; therefore, they do not have intelligence or a nonlinear modulation ability. As a result, the tags are expected to be printed on a paper substrate with metallic ink.

As chipless RFID technology is an intermediate between barcode and classical RFID, it combines the advantages of both technologies in industrial applications. Since chipless RFID tags can be directly printed on papers, they may have a realization cost comparable with that of barcode labels and can be read in a non-line-of-sight condition. At present, chipless RFID technology has no real industrial use apart from surface acoustic wave (SAW) devices. However, SAW tags are printed on piezoelectric substrates, which are costlier than papers, and hence they cannot be considered a real potential alternative to the barcode. Apart from SAW, chipless RFID is a new technology which was first discussed in 2002 by Richard Ribon Fletcher in his PhD dissertation at the Massachusetts Institute of Technology.

With the increasing interest in chipless RFID, many researchers have redirected their studies to this technology. Until now, increasing the coding capacity of chipless RFID tags, which is no more than 40 bits, has been an important factor. From the reader's point of view, the demand for this amelioration is crucial. Chipless RFID technology requires a reader that is cheap, compliant with UWB regulations, small and preferably integrated inside a smartphone such as that used for other PAN communication techniques starting from the barcode. Thus, this book helps to expound the UWB chipless RFID reader technology. It describes in detail the design of a reader for chipless RFID.

Chapter 1 is an introduction to RFID technology, focusing in greater detail on chipless RFID. This chapter further discusses the different characteristics of RFID, the barcode and chipless RFID technologies. It explains in detail the operating principle of UWB chipless tags together with their reading techniques.

Chapter 2 presents the state of the art of UWB chipless RFID reader technology. All the proposed readers in the literature are based on frequency-modulated continuous-wave (FMCW) and stepped-frequency continuous-wave (SFCW) approaches. These technologies are compared with impulse-radio UWB (IR-UWB) in terms of UWB regulations and reading time.

Chapter 3 introduces the design of a first IR-UWB-based chipless RFID reader. It is based on an equivalent time approach to reduce its realization cost while maintaining high acquisition performance.

Chapter 4 presents a second reader version which was developed with the aim of obtaining a short reading time and low sampling noise. After a theoretical introduction to sampling noise and quantization noise, this chapter discusses in detail the reader hardware architecture. Compared with the first version introduced in Chapter 3, this version allows for flexibility in terms of equivalent sampling rate, acquisition frame length and position.

Chapter 5 discusses the design of a low-jitter fully tunable UWB pulse generator as well as UWB planar antennas. The pulse generator can be used with the proposed readers in the emission stage and radiative elements to decrease the realization cost of the reader. This chapter also presents simulation and characterization results.

Chapter 6 presents a new reading technique based on polarization diversity. This approach can be used to read cross-polarized tags (i.e. chipless tags known for robust detection in a real environment) independently of their orientation. After a theoretical introduction, this approach is first validated with test equipment and then with the reader proposed in Chapter 4.

Marco GARBATI
Etienne PERRET
Romain SIRAGUSA
March 2018

# 1

# Introduction to Chipless RFID Technology

## 1.1. Introduction

In this chapter, chipless radio frequency identification (RFID) technology is introduced. After a brief discussion of classical RFID systems, we focus here on chipless RFID. These two technologies are compared with barcode technology. In this book, we focus on the reader part of the chipless technology based on ultra-wideband (UWB). Thus, the principle of operation of UWB chipless tags will be given in more detail. This chapter is organized as follows:

– Section 1.2 provides a general introduction to RFID technology and compares its main features with barcode technology.

– Section 1.3 presents chipless RFID technology in UWB. The principle of work of time-coded and frequency-coded tags is also given.

– Section 1.4 concludes the chapter.

## 1.2. Introduction to RFID

### 1.2.1. *Introduction*

RFID is among the commonly used technologies concerned with the identification of objects and animals (including humans). It is generally stated that the origin of RFID technology dates back to 1945, when the well-known Léon Theremin espionage equipment was used to retrieve conversations made inside the office of the US ambassador in Moscow. The "Thing" was totally passive and able to modulate its backscattered signal, once illuminated by an external electromagnetic source, according to the acoustic waves produced inside the ambassador office. At present, an RFID system retains the same principle of work except for the absence of the modulating acoustic waves, which are substituted with the smartness of an application-specific integrated circuit (ASIC) that performs the modulation.

During their long evolution, RFID systems have evolved in several directions due to the lack of standardization for many years and the huge number of distinct application requirements. To date, the most commonly used RFID systems may be classified according to their frequency band as: low frequency (LF), high frequency (HF), ultra-high frequency (UHF) and super-high frequency (SHF). Except for LF, all others use the unlicensed Industrial, Scientific and Medical (ISM) bands.

### 1.2.2. *Passive, semi-passive and active transponders*

Essentially, an RFID system is composed of a reader and a bunch of transponders inside the reading volume. The transponder may be passive, semi-passive or active. A passive transponder does not have a battery. Its chip is energized by harvesting from the reader transmission electromagnetic (EM) wave, i.e. reader interrogating signal. For most passive transponders, the transmission of information is performed in a passive way. Once energized, the chip is able to vary the load impedance connected to its antenna, generally between two different values. It modulates the amplitude of the impinging reader signal (down-link), which is backscattered toward the reader (up-link), ensuring communication between two actors. This is called load modulation, which is summarized in Figure 1.1.

The reader usually adopts either an amplitude shift keying (ASK) or a frequency shift keying (FSK) modulation with a dual objective: (1) to optimize the bandwidth available according to international regulations and (2) to maintain the transponder chip well energized to accomplish the communication. Conversely, the transponder can respond by modulating the amplitude and the phase of the reflected carrier wave by changing the chip impedance between two states. It can adopt modulation schemes with subcarrier generation such as binary phase shift keying (PSK), amplitude shift keying (ASK) and frequency shift keying (FSK). These schemes have a robust implementation at the expense of a reduced communication throughput. In a direct modulation scheme, each bit is associated with one reflecting carrier wave amplitude variation with a fixed bit period. It ensures the fastest communication at the expense of low noise immunity [MIC 02].

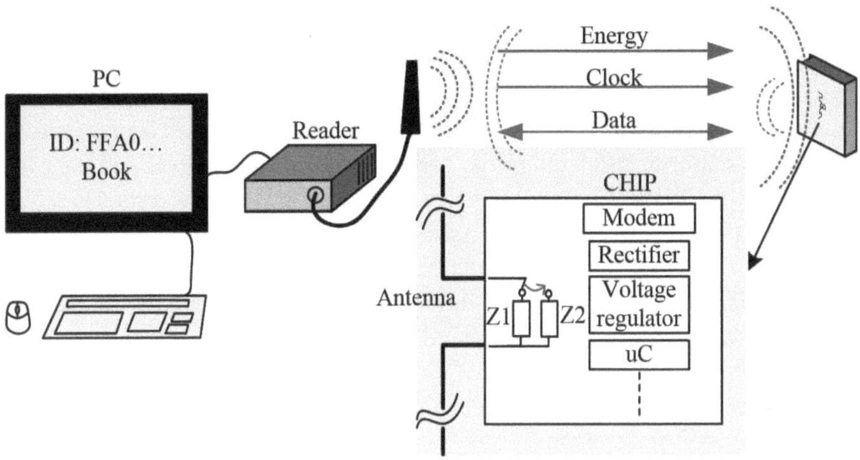


**Figure 1.1.** *Schematic of an RFID system, in which the reader interrogates the transponder that provides energy to wake up the ASIC (chip), clock for synchronization purposes and data as request code (down-link). Then, the chip replies by modulating the antenna load impedance between two values, providing a backscattering modulation to the reader (up-link). For a color version of this figure, see www.iste.co.uk/garbati/chipless.zip*

According to Figure 1.1, the transponder chip presents an analog part directly involved in the communication and a mixed part represented by a microcontroller. The analog part is formed of a modulator and demodulator (modem), a voltage rectifier to harvest the energy from the carrier wave and a voltage regulator to provide the correct voltage to the chip. The digital part is a hard-core microcontroller that may embed memories, inter-integrated circuit (IC2), serial peripheral interfaces (SPI) and analog sensor interfaces. The read range of a passive UHF transponder is lower than 10 m (short range) and limited by the harvester efficiency to about 20% [LE 08].

A semi-passive or active transponder is equipped with a battery to provide energy at the chip and eventually at peripherals. The communication retains the load modulation scheme, while the part of the chip in charge of the modulation is energized directly from the battery. The transponder can be woken up, in a cooperative way, from the battery and the harvested reader interrogating signal. Thus, the transponder is read at distances of up to 100 m for a UHF semi-passive transponder (moderate range). An active transponder has a radio frequency (RF) transmitter block used to communicate with readers as well as other transponders. Its read range is of several hundreds of meters (long range) [DOB 12].

RFID is a narrowband technology regulated by the European Telecommunications Standards Institute (ETSI) in Europe and by the Federal Communications Commission (FCC) in the USA. The LF and HF RFID applications and standards primarily use frequencies of 125 and 134 kHz and 6.78 and 13.56 MHz, respectively. The UHF interests 433.92 and 869 MHz in Europe (915 MHz in the USA), and 2.45 GHz, while the SHF is for 5.8 GHz. Several standards have been proposed from the physical to the software application level, where some are international such as the ISO/IEC and others are proprietary. The standardization helps the manufacturer to realize products with high level of interoperability. For instance, the ISO 18000 series is taking care of the air interface protocol (frequency, data modulation, etc.) for systems, to be used to track goods in the supply chain for RFID, in all the aforementioned frequency bands [VIOL 05]. Some of the currently used standards are shown in Figure 1.2 [SAT 12, DOB 05].

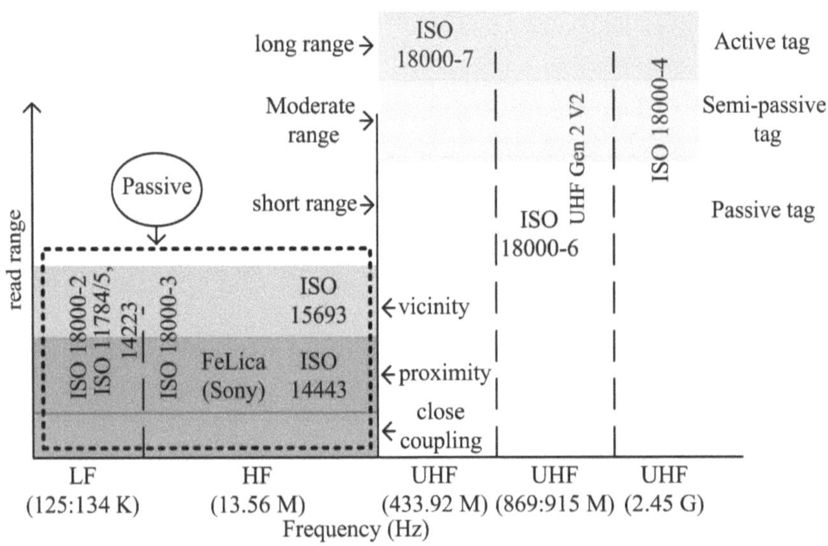


**Figure 1.2.** *Reduced collection of RFID standards. From international organizations (ISO, GS1) to private enterprises such as Sony. For a color version of this figure, see www.iste.co.uk/garbati/chipless.zip*

### 1.2.3. *RFID versus barcode*

Due to its minimal label cost, the barcode technology is used more than RFID in item identification. Nowadays, the realization cost of a barcode label is orders of magnitude lower than that of a passive RFID transponder. A barcode can be printed on paper substrate with non-conductive ink based on the principle of optical reflection. Due to the use of RF, a conductive ink is needed in RFID to generate surface currents. It is used in HF RFID to print out the antenna in label tag (transponder) for transport ticketing. It is also a subject of study for passive UHF tags for tracking item applications and for wearable electronics [BJO 16, RIZ 16]. The conductive ink is more expensive than non-conductive ink and, hence, an RFID tag cannot compete with a barcode label in terms of cost, where the cost of the chip can be prominent. The cost of an RFID transponder varies depending on its characteristics. It ranges from a few tens of euro-cents for inlay EPC GEN2 tags to hundreds of euros for sophisticated long-life active transponders. A barcode label may cost less than 0.5 euro-cents and even less if the barcode is directly printed on items.

A barcode label substrate may be composed of any solid material, and the label color should be as pale as possible. It is common to print the barcode out directly on the item surface to save substrate cost. For a printed UHF RFID transponder, the substrate can considerably affect the reading performance. Thus, it is hardly possible to print the transponder antenna directly on non-characterized tracking items. The relative permittivity of the substrate may vary the antenna operating frequency.

With the exception of cost, RFID technology offers many advantages compared with barcode. It is smart (chip) and has a higher read range and data bit capacity. It does not need a direct-line-of-sight to be read thanks to the use of RF, and hence the transponder can be hid and made invisible. It may also work in a harsh environment and be waterproof.

## 1.3. Chipless RFID

### 1.3.1. *Principle of operation*

Chipless RFID technology was developed by Richard Ribon Fletcher who published it in his PhD dissertation in 2002 [FLE 02]. Then, in 2005, Michael Pettus patented a chipless RFID system, where the tag was composed of a multitude of RF antennas, each providing different orientations and phase characteristics so as to encode the tag identification (ID). In this system, the reader scans an area and, using radar imaging techniques, can retrieve the tag information [PET 05]. Since then, chipless technology has been gaining a continuous growth of interest worldwide, mostly from universities. It is a technique of identification where the tag does not have any electronic component attached on its surface and relies on its EM property for the identification.

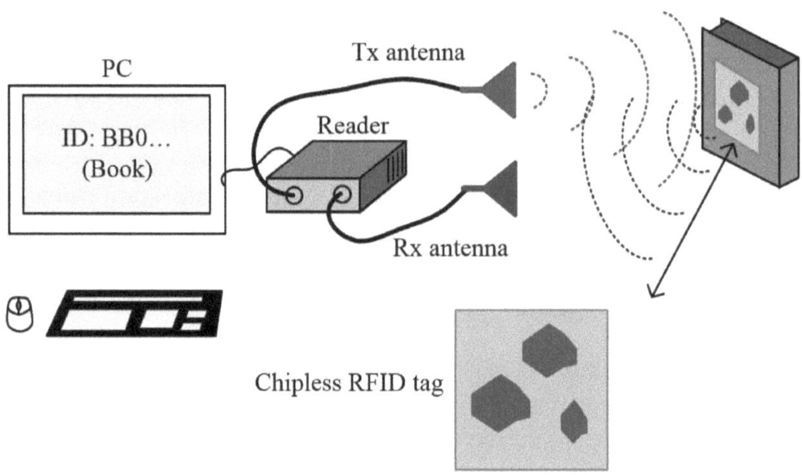


**Figure 1.3.** *Schematic of a chipless RFID system. The reader interrogates the transponder with a UWB signal, and then the backscattered signal from the tag is analyzed to retrieve the tag ID. For a color version of this figure, see www.iste.co.uk/garbati/chipless.zip*

A chipless RFID reading system is shown in Figure 1.3. The tag is commonly composed of a pattern (like in barcode) made up of conductive material, created upon a substrate with etching or printing technique. The principle of work is similar to that of a radar application, where non-radar cross section (RCS) modulation is expected. In contrast to RFID, the reader transmits a wideband signal to the environment, and the tag-backscattered signal is measured and analyzed, in time and frequency domain, to retrieve the desired information. The total absence of a chip is denied the opportunity of any signal modulation, and therefore chipless RFID technology is not compatible with narrowband identification utilization, such as classic RFID, and needs to be wideband.

A chipless RFID tag can be seen as a passive filter, where the information is coded based on its impulsive response: a larger available bandwidth means a higher coding capacity. A chipless RFID system needs to comply with UWB regulations in the 3.1–10.6 GHz band region. The UWB at larger frequency is reserved for different applications. Chipless technology is developing also in extremely high frequency [POP 16a] and terahertz [HAM 12], where a sort of two-dimensional (2D) imaging encoding technique may be used. However, the latter two frequency bands are not the subject of study in this book.

### 1.3.2. *Chipless RFID in-between RFID and the barcode*

Chipless RFID technology may serve as an intermediate between conventional RFID and the barcode. The absence of electronic devices with their package may represent a significant abatement of tag realization cost. Thus, a chipless RFID tag may be cheaper than an RFID tag and perhaps comparable to a barcode [VEN 13a]. To date, the cost comparison is a hard topic. Despite the absence of electronic devices, the label pattern may have a total surface area larger than that of a passive UHF RFID tag on an expensive substrate. Thus, most chipless tag prototypes have a realization cost higher than that of UHF RFID tag [PRE 10a]. However, studies on miniaturized single-layer chipless RFID tags for UWB compatible with ultra low-cost realization techniques are available in the literature. For instance, in [VEN 13a], chipless tags printed with conductive ink, using flexography, are presented. In [VEN 12a], the authors present a miniaturized chipless RFID tag of dimension 2 cm × 2 cm with 10 bits. The tag was realized on a rigid FR-4 substrate with etching technique and read with expensive laboratory equipment.

In contrast to the barcode, a chipless RFID tag provides the opportunity to be read without a direct line-of-sight. This is a significant advantage because the tag may be hid and read through objects. It opens up to plenty of applications for chipless RFID, where barcodes may not be used, such as discretion for luxury packaging and security. In contrast to a barcode, a chipless RFID tag may offer sensor capabilities [TED 13]. In [FEN 15], the authors present a printed chipless RFID tag on paper substrate designed as humidity sensor, with interdigital capacitor implementation. In [SCH 14], a time-domain-coded chipless temperature sensor was proposed, which was realized with screen-printing technology on an alumina substrate, with a central frequency of 7 GHz and a −10 dB bandwidth of 2 GHz.

The chipless read range is often considered higher than that of the barcode and lower than that of passive UHF RFID tags. This may be debatable, in fact for the chipless RFID tag, the read range is also correlated with its coding capacity in terms of the number of bits. Some chipless RFID tags, based on time-domain reflectometry (TDR), may have a read range over 1 m, but with a negligible number of bits, usually lower than 10. They are often realized on high-cost and RF-performing substrates [RAM 11]. Higher-bit-capacity chipless RFID tags have a reduced read range in the order of a few tens of centimeters.

The read range depends not only on the tag characteristics, but also on the reader performance and its power emission (UWB regulations). Studies often focus on read range, bit capacity and, in general, chipless RFID tags with high performance, but using expensive laboratory equipment as reader, in an anechoic environment and slighting taking into consideration the UWB regulations.

Multipath effects which may affect the reading capability are present in chipless RFID. Conversely, in the barcode, the interrogating optical reader signal can be well focused on the tag, which allows for reducing multipath effects. In addition, a barcode label needs to be read with direct line-of-sight, as the reading optical signal cannot go behind the label and eventual reflective perturbing signals are not present. Tables comparing the barcode, chipless RFID and UHF RFID in terms of tag performances and realization cost are commonly found in the literature [PER 14]. Almost all of them rank all the parameters well, except tag realization cost, for UHF RFID. Furthermore, the chipless RFID outperforms barcode in the reading range, with comparable realization cost and lower bit capacity. From our experience in the field to date, there has been no evidence supporting either the barcode or chipless RFID in terms of reading performances. However, today, the barcode is the winner in terms of tag realization cost, but the match between passive UHF RFID tags and chipless RFID tags is still open.

### 1.3.3. *UWB chipless RFID*

Chipless RFID technology has to comply with international regulations as with all electronic communication systems. A chipless RFID system works with UWB because the tag bit-capacity increases with the frequency bandwidth. UWB technology is applied to systems that use large bandwidths and low power to send or retrieve information. The American FCC regulation concerning UWB is in Part 15 of Title 47 [FCC 17] and grants the band 3.1–10.6 GHz for communication systems in indoor applications such as chipless RFID. The ETSI regulation, in terms of UWB systems, is stricter than FCC. It allows for the same band of FCC, but with lower power between the bands 6 and 8.5 GHz [ETS 04, ETS 10]. This book concerns the readers for UWB chipless RFID tags that may be classified as time- and frequency-coded tags.

#### 1.3.3.1. *UWB time-coded chipless RFID tag*

A UWB time-coded chipless RFID tag, in its most simple representation, is composed of one UWB antenna connected to a transmission line, along which some impedance mismatchings are created. A schematic example of a four-reflection-based tag is shown in Figure 1.4(a). Once interrogated with a UWB reader, the tag reflected signal may be analyzed in the time domain and delays between the different reflections can be estimated. The values of these delays encode the tag information. The tag response in time is composed of a tag structural mode and an antenna or tag mode. The former is due to the specular reflection of the tag assemblage, and the latter is the tag behavior, where the antenna is transferring energy to the transmission line, and all the reflections due to impedance mismatchings occur. The antenna (tag) mode is the essential information in chipless technology, and is exploited in a copious manner with signal processing. In the case of a time-coded chipless tag, the structural mode is also useful. It gives information about the tag distance and is also exploited to calculate the relative delays of the reflections to decode the tag ID. An example of an encoding technique is shown in Figure 1.4(b), corresponding to the tag in (a). The time domain is divided into four areas, and each is associated with one of the four signal reflections that constitute the tag mode. On the basis of the position of the reflection inside the area, a different binary code is assigned. In the figure, the tag ID is $00001001_2$, which corresponds to eight bits of information.

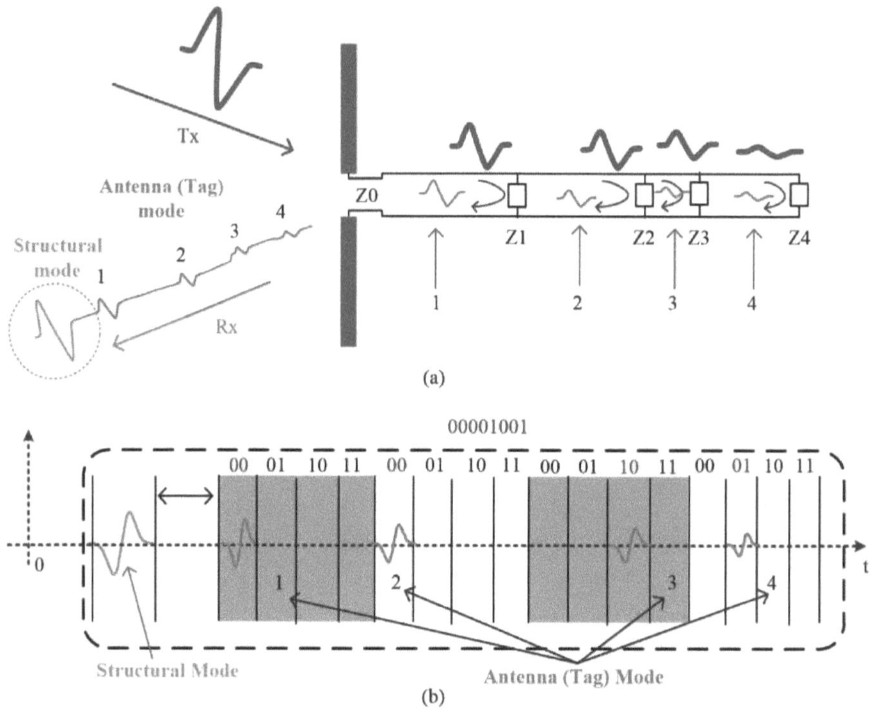


**Figure 1.4.** *a) Schematic of a UWB chipless RFID tag based on a time-coded approach. The tag has four impedance mismatchings (Zi) that originate four main reflections toward the reader. b) The tag information is encoded with the temporal distances between the structural mode and the various reflections in the antenna (tag) mode. For a color version of this figure, see www.iste.co.uk/garbati/chipless.zip*

Figure 1.4 is overly simplified because the impedances $Z_i$ generate multiple reflections. For instance, the first reflected signal by $Z_2$, before reaching the antenna, will be in part reflected back by the impedance $Z_1$ and so on. These effects will create disturbances in the reader receiving signal, and the low-energy reflected signals will be decoded with much difficulty.

The major drawback of a time-coded tag is represented by the transmission line length that may be substantial. Considering a Gaussian monocycle interrogating signal with a pulse full width of about 500 ps, the minimum registered distance between reflections in the tag response should be at least 500 ps, which corresponds to 250 ps transmission line delay between two impedances (2 ways). If the tag transmission line has a relative permittivity $\varepsilon_r$ of about 4.4 (FR-4), then the wave speed inside the line $v_L$ is approximately

$$v_L = \frac{v_0}{\sqrt{\varepsilon_r}}. \qquad [1.1]$$

where $v_0$ is the speed of light in a vacuum environment, which is about $3 \times 10^8$ m/s. In such a configuration, $v_L$ is about $1.43 \times 10^8$ m/s, and multiplying it by 250 ps returns to a minimum distance between two reflectors of 3.5 cm. Hence, a UWB time-coded chipless RFID tag may have a significant dimension. It can be reduced using a smaller number of reflective elements (lower bit capacity), higher permittivity materials and bending techniques.

In a chipless RFID tag, all the impedances $Z_i$ have to be realized with layout techniques (chipless). To reduce as much as possible the loss of the transmission line, the impedance elements $Z_i$ should have zero resistivity. This leads to the realization of pure imaginary discontinuities. However, they are known to be narrowband in its simpler realization (stub). Consequently, in the literature, it is possible to find UWB time-coded chipless RFID tags, which are not really *chipless* because the impedances are realized with discrete components [ZHA 06, SCH 09].

The use of surface acoustic wave (SAW) technology is not recent and is used in the field of electronics in various applications, such as filters, oscillators and sensors. They are based on acoustic waves instead of electromagnetic waves, whose speed in free space is about $10^5$ times lower than the speed of light. This feature can be exploited to create much smaller RF devices. A SAW component needs two interdigitated transducers (IDT) and a piezoelectric substrate that can transduce between electrical and mechanical energies. Time-coded chipless RFID SAW tags are a reality nowadays and are produced by several enterprises. These tags are working in the ISM band at 2.45 GHz (narrowband). The bit capacity may exceed 128 bits with a tag dimension of about 10 mm for side not considering the antenna [HAR 02]. The major drawback of a SAW tag is the realization cost caused by the need of piezoelectric substrates as well as by the use of etching techniques for metallic layer deposition. It may not represent a competitor either for the barcode or for UHF RFID passive technology.

In 2008, a patent on the use of cellulose substrate to propagate acoustic wave was deposited [KIM 08]. The tag was characterized by the presence of an antenna, a SAW-based IDT and a several reflectors for reflecting the surface acoustic wave. All were mounted on cellulose substrate that would help to reduce the tag substrate cost. However, there are other challenges, such as the printability of the tag and the immunity against the environmental condition.

##### 1.3.3.1.1. UWB time-coded chipless RFID tag with single-line discontinuity

A UWB time-coded chipless RFID tag with only one discontinuity line can be readily designed. It needs to have either a zero impedance (short circuit) or infinite impedance (open circuit) to reflect as much energy as possible in the direction of the antenna. It is simple to realize a large band open circuit with layout technique while, for a short circuit, it may depend on the transmission line technology. In a microstrip, a short circuit needs a via, while for a coplanar it does not. This approach has limitations in coding capacity, but is beneficial in terms of read range and dimensions. The principle schematic of a single-reflector UWB time-coded chipless RFID tag and a reading process with a bi-static reader are shown in Figure 1.5.

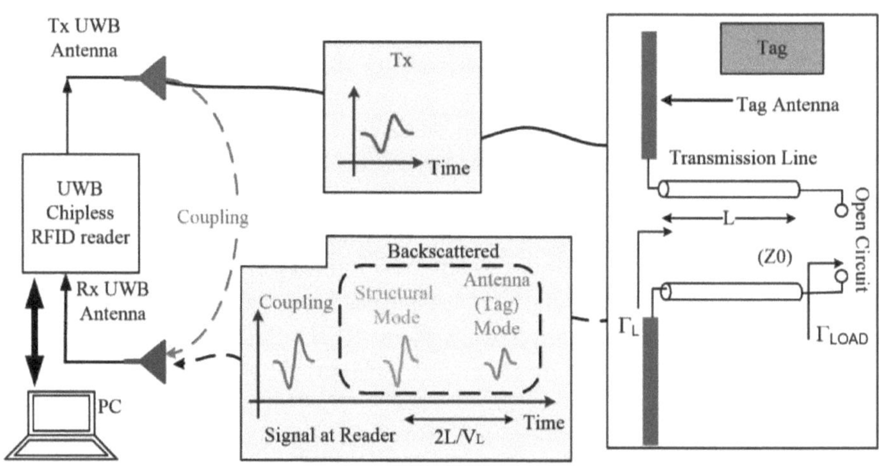


**Figure 1.5.** *Schematic of a UWB chipless RFID tag based on a time-coded approach using only one line discontinuity (open-circuit load). A bi-static UWB chipless RFID reader is used. For a color version of this figure, see www.iste.co.uk/garbati/chipless.zip*

For a bi-static configuration, the coupling between the two reader antennas is reduced (green dashed line in Figure 1.5) compared to a mono-static configuration. However, it is omnipresent in short-range radar-based applications such as chipless tags. If the coupling of the reader antennas and the tag are not influencing each other, the coupling effect can be removed from one tag identification process in post-processing. Two measurements are necessary, one without the tag and another with the tag (background subtraction [RAM 11]).

A typical bi-static co-polarization measurement result of a time-coded chipless tag at several distances up to 2 m in a practical environment [RAM 12a] is shown in Figure 1.6. It uses known detection methods (background subtraction and continuous wavelet transform (CWT)) [LAZ 11]. As can be observed, both the structural and antenna (tag) modes can be detected. The tag layout along with its measured and simulated reflection coefficients is shown in Figure 1.7. The tag consists of a microstrip-fed UWB monopole [WHY 08] fabricated on a Rogers RO4003C substrate.

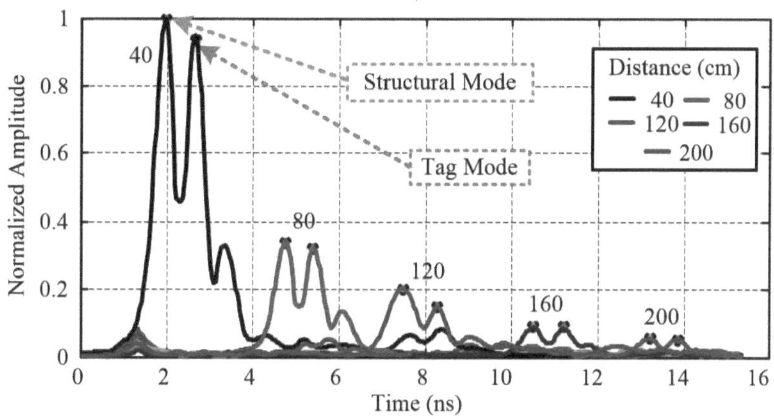


**Figure 1.6.** *Normalized result measurement after background subtraction (top) and CWT for the monopole tag at several tag–reader distances in a practical environment. Original figure can be seen in [RAM 12a]. For a color version of this figure, see www.iste.co.uk/garbati/chipless.zip*

The obtained read range of 2 m is remarkable, as well as the reduced dimension of the tag (5.3 cm × 4.3 cm), making it smaller than a credit card, which is 8.560 cm × 5.398 cm according to the ISO/IEC 7810 ID-1. The tag information capability depends also on the reader performance and post-processing power. A value lower than 10 bits is reasonable and, hence, this kind of UWB chipless RFID tag is more suitable for sensing and localization applications rather than identification [RAM 11, RAM 13].

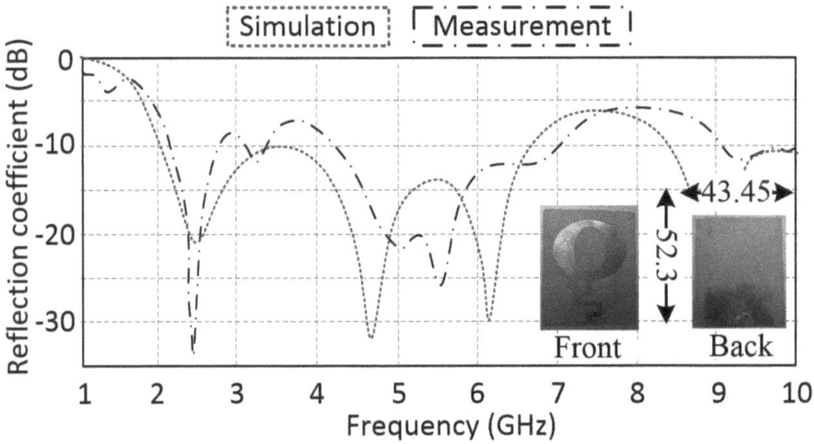


**Figure 1.7.** *Layout of the microstrip-fed UWB monopole tag with dimensions in millimeters and its measured and simulated tag reflection coefficient. Original figure can be seen in [RAM 12a]. For a color version of this figure, see www.iste.co.uk/garbati/chipless.zip*

### 1.3.3.2. *UWB frequency-coded chipless RFID tag*

A UWB frequency-coded chipless RFID tag uses an approach based on its spectral response to encode information. The tag geometry is modified to account for specific amplitude and phase characteristics over the frequency band of interest. A well-known structure of such a tag is composed of two UWB antennas (one for transmission and the other for reception) and, between them, a passive filter is realized with distributed elements, as shown in Figure 1.8(a).

If the passive distributed filter shown in Figure 1.8 is considered lossless, part of its input signal will be reflected and part transmitted to the other antenna. The antennas are linear-polarized with a cross-polarization configuration (in a perpendicular direction). If a reading approach similar to that shown in Figure 1.5 (UWB bi-static) is used to read the tag, the reader antennas have to be lined up with the tag antennas, to read the tag successfully. The cross-polarization configuration allows for measuring only the part of the UWB reader interrogating signal that crosses the tag filter, and therefore the part of the signal where the tag ID is encoded.

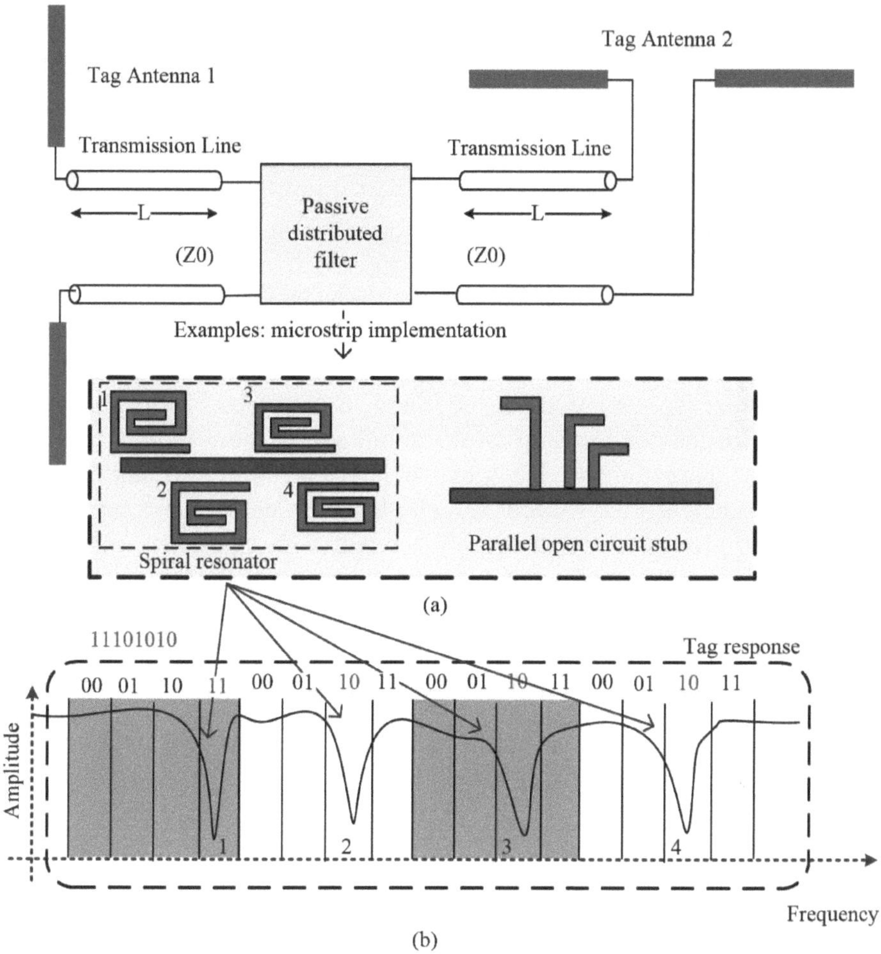


**Figure 1.8.** *Schematic of a frequency-coded UWB chipless RFID tag based on cross-polarization. Two examples of passive distributed filters are shown, one based on spiral resonators [PRE 09] and the other based on parallel open-circuit stubs [NIJ 12]. For a color version of this figure, see www.iste.co.uk/garbati/chipless.zip*

Figure 1.8(a) shows two examples of passive filters compatible with microstrip technology: one based on notch spiral resonators and the other based on parallel open-circuit stubs. The first solution applied to chipless technology was proposed in [PRE 09]. Spiral filters were placed along with the microstrip line, and each resonator was introduced with a diverse band-stop resonance. It was designed in the band 3.1–7 GHz and realized in Taconic TLX-0 with dimensions of 8.8 cm × 6.5 cm. The information is encoded in the amplitude part of the spectrum reaching 35 bit of information. When varying the dimension of the spirals, their first-mode resonant frequency will change accordingly, as will the filter characteristics.

The solution shown in Figure 1.8(a) using parallel open-circuit stubs was proposed in [NIJ 12]. The tag has a limited 8-bit capacity, and the coding still depends on the spectrum amplitude. The antennas are two cross-polarized monopoles, and the tag measures 5.0 cm × 3.0 cm, without considering the antennas. The bandwidth of interest is between 2 and 4 GHz, and the substrate is of high performance from Rogers. In [NAI 16], the authors presented a UWB frequency domain chipless RFID tag, where only the filter section was implemented and characterized with laboratory equipment. Instead of focusing on the amplitude of the response, they took care of the phase through the group delay. The filter was composed of cascaded commensurate C-sections, whose length determine the frequency with higher group delay variation. To retain a small filter dimension, a folding approach is proposed. As yet, the proposed architecture of a UWB frequency-coded chipless RFID tag presents a discrete bit capacity of approximately 40 bits. On the contrary, it expects a reader in bi-static configuration, with two cross-polarized antennas, and the tag needs to be aligned with the reader antennas for maximum performance.

#### 1.3.3.2.1. Join antennas and filter

The frequency-coded chipless RFID tags proposed in [PRE 09, NIJ 12] are based on spectrum amplitude modulation, using a set of independent notch filters, in the absence/presence of an encoding sort (see Figure 1.8). An exhaustive overview of UWB chipless encoding techniques is given in [TED 12].

Using the RF encoding particle (REP) approach introduced in [PER 14, VEN 16], it is possible to obtain the requested spectrum amplitude modulation of the tag in a smaller structure [VEN 11, VEN 12a]. This structure is composed of several resonant scatterers (particles), in which the two antennas and the filter are enclosed. For instance, the tag proposed in [NIJ 12], which is composed of parallel open-circuit stubs used as a band-stop filter (BSF), can be translated in an equivalent tag based on an

REP approach. It is shown in Figure 1.9, in which only three resonators of [NIJ 12] are shown for simplicity.

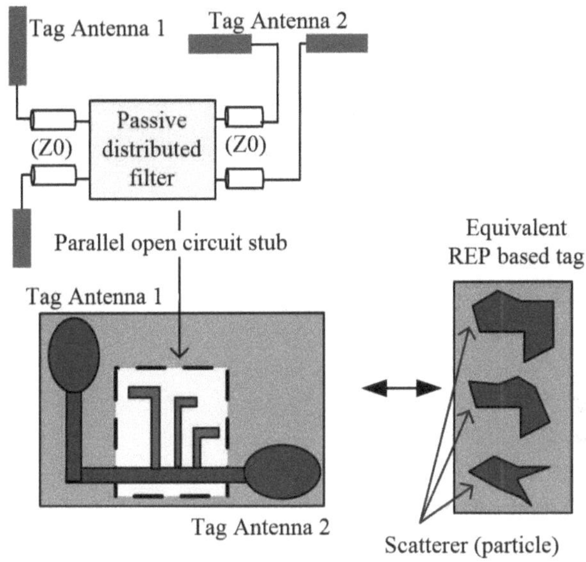


**Figure 1.9.** *Schematic of a UWB frequency-coded tag based on microstrip with two UWB disk monopole antennas and three parallel open-circuit stubs similar to [NIJ 12]. The tag can be realized with three resonant scatterers using the REP approach, with similar performance to the original tag. For a color version of this figure, see www.iste.co.uk/garbati/chipless.zip*

The REP approach is a complex subject; from a theoretical point of view, one interest is that a REP chipless tag can be considered as a radar target. The specificity here is that the tag (target) is designed to have a known number of resonant frequencies. The singularity expansion method shows that these frequencies, which are related to the tag mode, are independent of the reading method. Thus, they can be used to encode the information [BAU 91, REZ 13].

The study of the behavior of a loaded antenna, constructed with linear reciprocal materials, is fundamental to a better understanding of the complexity of REP using both analytical and physical approaches. In [GRE 63], the author presents a study about the prediction of the backscattering property of a loaded antenna with some minimal measurements. Once excited with an impinging electromagnetic field, two different scattering processes occur, one load-independent and the other fully dependent. As previously explained, the first process is due to the immediate reflection that occurs on the metallic part of the antenna, through the formation of surface currents (structural mode scattering). Instead, the second process concerns the radiative property of the antenna connected at that specific load (antenna mode scattering). This can be seen as the property of the antenna to convey the incoming electromagnetic field toward the load, which is responsible for its dissipation. The tag of Figure 1.9 with the REP approach is composed of three scatterers. Each scatterer has its structural and antenna (tag) mode scattering, and thus can be seen as an antenna connected to a complex load.

One of the first uses of the REP approach to chipless RFID was in [JAL 05], where the authors realized a barcode-type chipless tag composed of five scatterers. Each scatterer was considered as a short-circuited dipole antenna and not yet, strictly speaking, as an REP scatterer, a relatively new concept [PER 4]. Each scatterer was composed of a strip with different lengths, in microstrip technology. It was designed in the bandwidth 5.4–6 GHz on Taconics TLY-5 and shown to have a five-bit capacity and a read range of tens of centimeters. The measurement was performed with a vector network analyzer (VNA) in co-polarization, and it shows a band rejection effect of the tag at the strip first resonance frequency ($\lambda/2$). This was due to the destructive effect between the scatterer structural mode and the antenna (tag) mode.

Since [JAL 05], remarkable progress has been made on this approach for chipless RFID and many works have been published by laboratories worldwide. The bit capacity of this type of tag has been increased using more elaborated scattering particles, with magnitude as well as phase coding information, and using mainly a cross-polarization reading approach with a more elaborate post-processing treatment. In [VEN 11], a tag based on C-like resonators (scatterers) was presented, which was realized on one metallic layer in a low-cost FR-4 substrate with a permittivity of 4.6. Using a coding based on the frequency resonant position and the phase characteristic of C-like resonators, the coding capacity was increased up to 22.9 bits, with a tag size of 2 cm × 4 cm, where the tag bandwidth was 2–7 GHz. The use of only one layer represents a significant reduction of the tag development cost, with the possibility of realizing the tag by using printing techniques.

The reading process in an anechoic chamber and in cross-polarization of a depolarizing REP tag introduced in [VEN 13b] is shown in Figure 1.10(a).

The tag was realized in Roger R04003 with a ground plane, which was based on eight resonators, each composed of five coupled dipoles in short circuit with the first resonant mode in the bandwidth of 3–7 GHz. Agilent N9918A was used as VNA with an emission power of 0 dBm. The dual-access Satimo QH2000 was employed, and the tag was placed at 15 cm. The detection process used was background subtraction. The measurement results are shown in Figure 1.10(b), where the eight resonant frequencies correspond to the eight visible peaks. The coding is based on the position of the resonances.

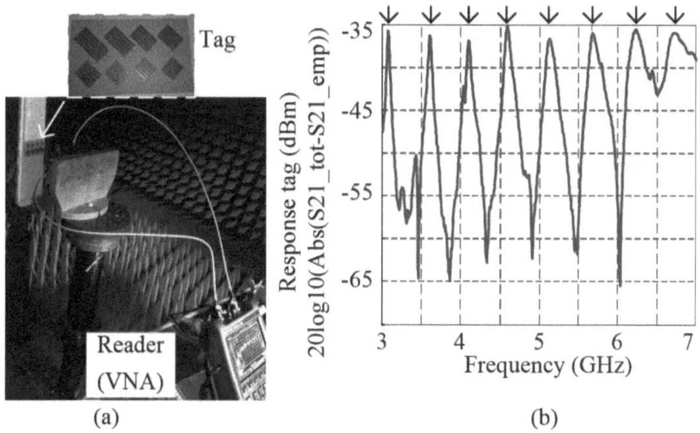


**Figure 1.10.** *Measurement of cross-polarization of a frequency-coded UWB REP chipless RFID tag proposed in [VEN 13b], in an anechoic chamber. a) Photograph of the test bench and the measured tag. Agilent N9918A is used to measure the S-parameter (S21), with an averaging factor of 10. The tag is placed at 15 cm from the dual-access dual-polarization Satimo QH2000 antenna. b) Tag response, with the positions of the eight resonant frequencies underlined*

## 1.4. Conclusion

In this chapter, chipless RFID technology based on the UWB was introduced. Chipless RFID may serve as an intermediate stage between more advanced RFID and the barcode. It may represent a cheaper identification solution than RFID, preserving non-direct-line-of-sight in the reading process and the possibility to embed sensor capabilities. These two characteristics represent significant advantages compared to barcode. On the contrary, a UWB chipless tag, in comparison with barcode, has lower bit capacity, lower multipath immunity and higher realization cost. The latter is mostly due to the cost of conductive ink in the case of printable tags. A

UWB chipless tag may be classified as time-coded and frequency-coded, where the latter uses REP technology to reduce its dimension.

Chipless RFID technology is based on a radar approach, where the backscattering energy from an object through its RCS is the heart of all the radar applications. Thus, the performance of the reader has an important impact on the reading capability. Parameters such as the read range, the impact of different tag orientations and tag bit capacity significantly depend on the reader. In the following chapter, the state of the art of chipless reading technology will be reviewed.

# 2

# UWB Chipless RFID Reader: State of the Art

## 2.1. Introduction

Chapter 1 introduced chipless RFID technology based on UWB. It highlighted its main characteristics by comparing with classical RFID and the barcode. The aim of this book is to describe the development of low-cost readers for UWB chipless RFID tags, which maintain a reduced reading time in accordance with international regulations. This chapter introduces the current state of the art for chipless readers that operate in the UWB ranging from 3.1 to 10.6 GHz.

All the proposed readers in the literature are based on a frequency domain approach, either a stepped-frequency continuous-wave (SFCW) or a frequency-modulated continuous-wave (FMCW) approach. Both approaches provide a transmitting signal sweep between the tag bandwidths to retrieve the desired information. An SFCW reader has a longer sweep time than a FMCW reader, where the transmitted CW instantaneous frequency varies with a staircase function, as shown in Figure 2.1(a). The transmitting and receiving instantaneous frequencies, f'(Tx) and f'(Rx), respectively, have a small offset, without generating any beat frequency. This approach is used in a commercial vector network analyzer (VNA) which is suitable for reading both frequency- and time-based chipless RFID tags. An FMCW approach is typically used in radar applications where a beat frequency is generated due to the offset $\Delta f$ between f'(Tx) and f'(Rx), as shown in Figure 2.1(b). This approach allows for a short sweep time compared with the SFCW approach, and both time- and frequency-based chipless RFID tag topologies can be read. However, a frequency-coded tag with FMCW is read with low accuracy, and thus requires a complex post-processing algorithm that may involve the Hilbert transform.

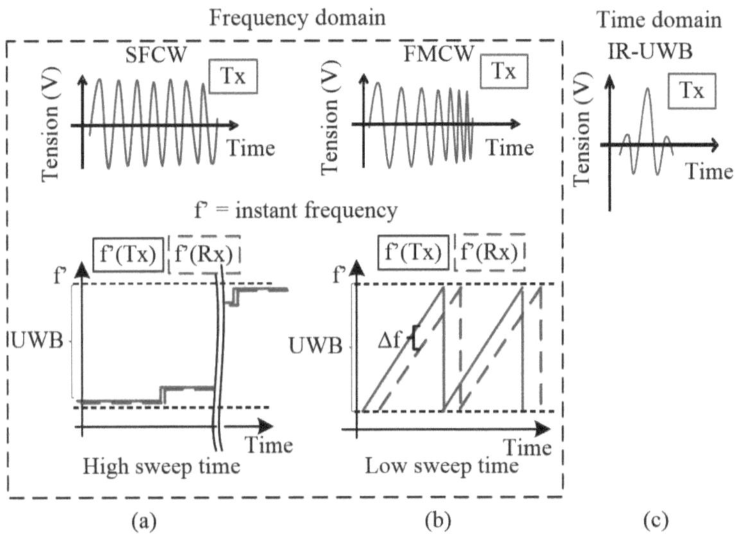


**Figure 2.1.** *a) Transmitting and receiving signals of a UWB chipless RFID reader based on the SFCW approach. It has a long sweep time, which is ideal for reading a frequency-coded chipless tag. b) Transmitting and receiving signals of a UWB chipless RFID reader based on the FMCW approach. It has a short sweep time, which is mostly used to read a time-coded chipless tag. c) Transmitting signal of a UWB chipless RFID reader based on the IR-UWB approach. For a color version of this figure, see www.iste.co.uk/garbati/chipless.zip*

This chapter is organized as follows:

– section 2.2 introduces the architecture of two SFCW readers, characterized by two mixers in reception, and an IQ demodulation scheme;

– section 2.3 details the readers based on the FMCW architecture and demonstrates their capability of reading frequency-based tags;

– section 2.4 introduces a reading technique based on the impulse-radio UWB (IR-UWB) approach, as shown in Figure 2.1(c). It represents a time domain approach, and until now, no IR-UWB chipless RFID readers have been reported in the literature. This section further provides a comparison between SFCW, FMCW and IR-UWB architectures in accordance with

UWB regulations. Finally, it proposes a link budget involving the three different approaches, with a typical frequency-coded tag considered a target;

– section 2.5 draws the conclusion of this chapter.

## 2.2. SFCW approach

A block schematic of a bi-static SFCW reader is shown in Figure 2.2. In the transmission mode, a phase-locked loop (PLL) is used to generate the CW-interrogating signal whose instant frequency f'(Tx) is controlled by digital and control units. In the reception mode, an I/Q demodulation scheme is used to recover the amplitude and phase difference between the reader's transmitting and receiving signals. The generation of a beat frequency is not required for this solution, and the sweep time of the PLL can be kept as high as desired. However, the sweep time should not be excessive to allow undesirable reading time. With this approach, it is possible to characterize the tag's impulsive response in the band of interest, and thus decode both chipless RFID tag topologies.

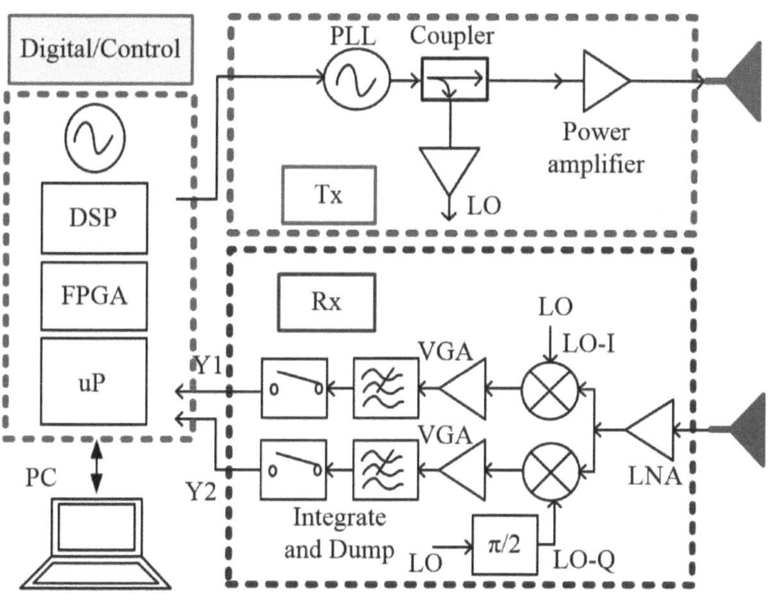


**Figure 2.2.** *A UWB chipless RFID reader based on the FMCW approach and I/Q demodulation. A beat frequency is not required, thanks to the use of two mixers*

A reader for UWB chipless RFID tags based on the SFCW approach was proposed in [PRE 10b]. It was designed to operate between 5 and 9 GHz, and a wideband yttrium iron garnet (YIG) sphere voltage-controlled oscillator (VCO) was used in transmission. The block schematic of the reader in [PRE 10b] is shown in Figure 2.3. It is based on a heterodyne configuration that generates an intermediate frequency (IF) signal to reduce the effect of flicker noise.

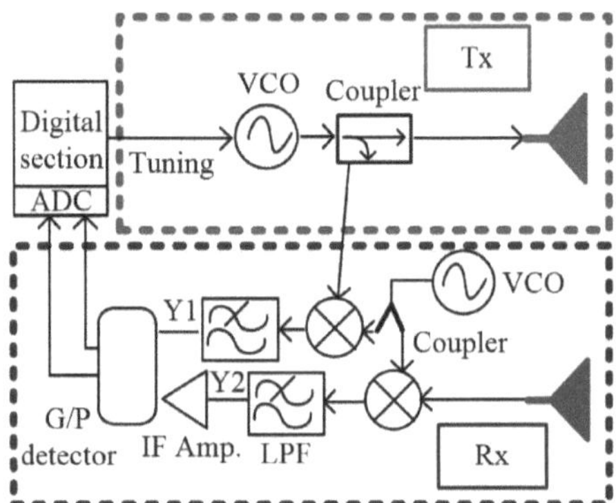


**Figure 2.3.** *Schematic of a UWB chipless RFID reader proposed in [PRE 10b]. It is based on a heterodyne configuration with the SFCW approach to reduce the effect of flicker noise due to active devices*

Two outputs of low-pass filters (LPFs), $Y_1$ and $Y_2$, pass through the AD8302 from analog devices. The AD8302 is a gain and phase detector for measuring the phase and amplitude difference between two signals. Since the amplitude of $Y_2$ depends on the amplitude of the tag response, an IF amplifier is required after the LPF. This is to compensate for the higher power present at the output of the coupler, compared with the input signal of the reader, where the tag response is present. The bandwidth of the reader is between 5 and 9 GHz, and its hardware cost without the antennas was 4,500 € in 2010. It is tested in an anechoic environment with an 8-bit frequency-coded tag at a few tens of centimeters. The emitted VCO output power is 16 dBm.

## 2.3. FMCW approach

This section introduces the main FMCW-based UWB chipless RFID readers proposed in the literature. The proposed readers have been designed using a specific tag's families, from frequency-coded [KOS 12] to time-coded [POP 16b]. Both readers have the same architecture and use different components that allow diverse reader performances.

In an FMCW approach, the instantaneous frequency of the transmitting and receiving signals has an offset $\Delta f$ which is proportional to the distance of the tag. This generates a beat signal in reception where the tag ID is encoded, as demonstrated in section 2.3.1.1 [CHA 00].

## 2.3.1. *Frequency-coded tag reader*

In [KOS 12], the authors proposed a UWB chipless RFID reader based on a chirp interrogation signal. This reader was optimized for frequency-coded tags, and its hardware architecture can be considered invariant with respect to a bi-static UWB FMCW radar. A photograph of the proposed reader with the tag used as a proof of concept and its schematic representation are shown in Figure 2.4. The reader has two boards: one for digital processing and the other (top) for analog circuitry. The analog board is made on Taconic TLX-0, which has a relative permittivity of 2.45 and a thickness of 0.5 mm. The reader operates in the bandwidth of 4–8 GHz, and is used with two UWB monopole antennas, with a bandwidth of 3.6–7 GHz. The tag is frequency-coded with 9 bits, which consists of two cross-polarized UWB antennas and spiral resonators operating as notch filters. The coding technique is based on a presence/absence approach.

### 2.3.1.1. *Limitations on reading a frequency-coded tag*

An analytical analysis is proposed to emphasize the limitations of a UWB FMCW chipless RFID reader on reading a frequency-coded tag. Equations are partially based on [KOS 12] and [CHA 00]. The tag must interrogate with a linear frequency-modulated signal, as shown in Figure 2.1(b). The interrogation starts at t = 0 s from the minimum frequency of the VCO, denoted as $f_0$, to the maximum frequency that is equal to $f_0 + BW$. The VCO is modulated with a sawtooth signal of repetition frequency $f_p$. The transmitted signal (VCO output) $X_t(t)$ toward the tag can be approximated using equation [2.1]:

$$X_t(t) = \cos\left(2\pi \int_0^t (f_0 + BW f_p t')dt'\right) \quad [2.1]$$

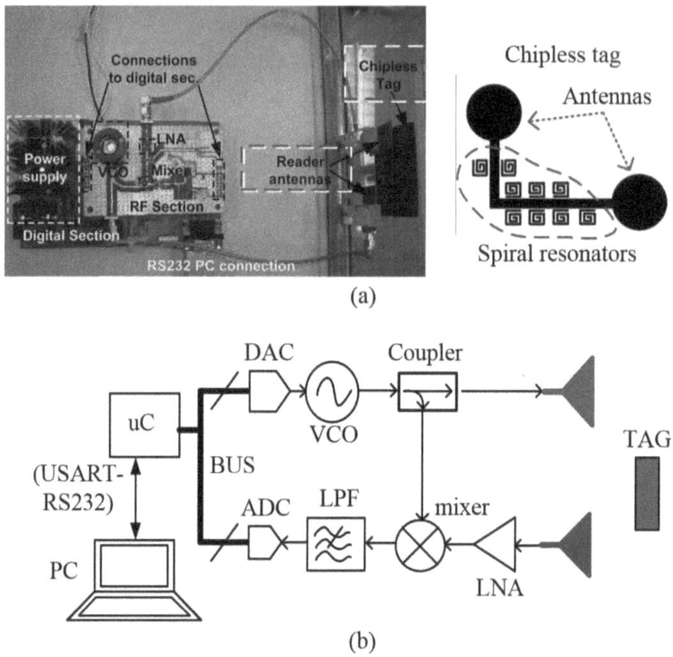


**Figure 2.4.** *a) Photograph of a UWB chipless RFID reader, as proposed in [KOS 12], with the frequency-coded tag used as a proof of concept. b) Architecture of the reader, which has the same hardware structure as that of an FMCW radar in a bi-static configuration. For a color version of this figure, see www.iste.co.uk/garbati/chipless.zip*

For the sake of simplicity, the four antennas, two readers and two tags are considered to be ideal irradiators, where the signal is only attenuated. $T_{path}$ is the time taken by the VCO output signal to reach the RF input of the receiving mixer via the tag, the antennas, the RF cables, the PCB components and the free space. This signal will vary in amplitude, $TM(f)$, and phase, $T\theta(f)$, with respect to the tag's transfer function. The objective of the reading process is to obtain the amplitude, $TM(f)$, and the phase, $T\theta(f)$, of the tag's transfer function. In the proposed case, the instant frequency of $X_t(t)$ is not constant as in the case of CW; therefore, in the first approximation, $TM(f)$ and $T\theta(f)$ can be substituted in the time domain expression of equation [2.1], which yields equation [2.2]:

$$Y_r(t) = TM(f^i)cos\left(2\pi \int_0^{t-T_{path}}(f_0 + BWf_pt')dt' - T\theta(f^i)\right) \quad [2.2]$$

where $Y_r(t)$ is the signal at the RF input of the mixer and $f^i$ is the instantaneous frequency of the transmitting signal at time $t - T_{path}$. The output of the mixer after the LPF can be expressed as follows:

$$Y_{ADC}(t) = TM(f^i)cos\left(\left(2\pi\int_0^t (f_0 + BWf_pt')\right)dt' - 2\pi\int_0^{t-T_{path}} (f_0 + BWf_pt')dt' - T\theta(f^i)\right)$$

$$= TM(f^i)cos\left(\left(2\pi\int_{t-T_{path}}^{t} (f_0 + BWf_pt')\right)dt' - T\theta(f^i)\right)$$

$$= TM(f^i)cos\left(\left(2\pi(f_0T_{path} + \frac{f_pT_{path}{}^2BW}{2} + f_pT_{path}BWt\right) - T\theta(f^i)\right)$$

$$\approx TM(f^i)cos\left(\left(2\pi(f_0T_{path} + f_pT_{path}BWt\right) - T\theta(f^i)\right) \qquad [2.3]$$

The instantaneous frequency of the transmitting signal is equal to $f_0 + BWf_pt$ (from equation [2.1]). Once $Y_{ADC}(t)$ is digitized, it is possible to obtain the two components $TM(f^i)$ and $T\theta(f^i)$ separately using the Hilbert transform. This is possible due to the generation of the beat frequency $f_pT_{path}BW$ which is equal to Δf, as shown in Figure 2.1(b). Equation [2.3] is the result of many approximations. The influence of the antennas and the coupling and echoes from the environment are not considered.

### 2.3.1.2. *Influence of the VCO sweep time on reading performance*

The accuracy that can be reached using an FMCW reader also depends on the product $BWf_p$. A small value of this product means the chirp interrogation signal slowly sweeps around the bandwidth of the tag. The tag output will reach its stable state according to the input instantaneous frequency. In comparison, the beat signal frequency $(f_pT_{path}BW)$ will be extremely small, thus making separate estimation of the amplitude and the phase components $(TM(f^i), T\theta(f^i))$ complex.

An increasing value of $BWf_p$ means a VCO with a higher input bandwidth and thus a reduced sweep time. This will in turn increase the beat frequency $(f_pT_{path}BW)$ and make the application of the Hilbert transform more effective. However, the approximation made in the time domain, with the introduction of the components $TM(f^i)$ and $T\theta(f^i)$, will no longer be of much use. Thus, the measured signals $TM(f^i)$ and $T\theta(f^i)$ will be in a larger bandwidth range and will not have a precise input instantaneous frequency, resulting in a less accurate estimation of the tag spectral characteristic.

### 2.3.1.3. *Simulation results*

This section considers the effect of a variable sweep time on the accuracy of reading a frequency-coded tag. A Matlab® Simulink model of an FMCW UWB chipless RFID reader is shown in Figure 2.5. It is used to read a tag based on one resonator in cross-polarization. All the components are considered to be ideal, and the attenuation of reader cables, antennas, free space loss and tag antennas is not taken into account. The propagation delay between the VCO output in transmission and the mixer's RF input in reception was modeled with two ideal transport delay blocks. The tag was interrogated with a CW signal whose instantaneous frequency linearly varied between 2 and 6 GHz, with a sawtooth modulating signal. Different simulations were performed by varying the sweep time $(1/f_p)$ between 100 ns and 100 µs, where the $BW$ was fixed at 4 GHz according to [KOS 12]. The tag resonator was modeled with an RLC series circuit terminated with a 50 Ω load. Its resonant frequency was 4 GHz with a quality factor of 100, which is consistent with REP-based resonant particles [VEN 16].

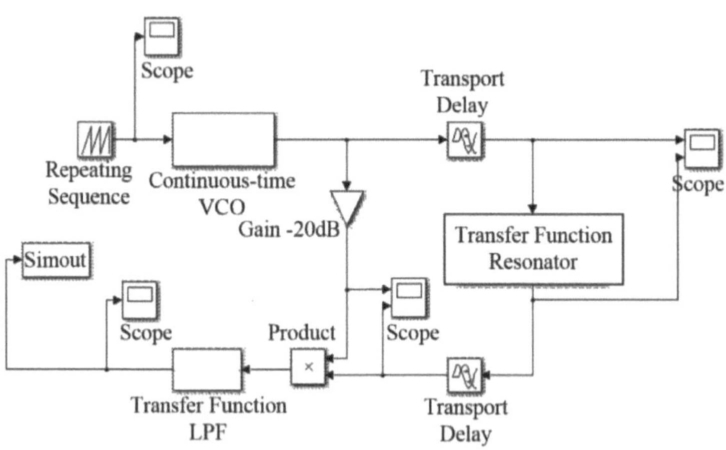


**Figure 2.5.** *Matlab® Simulink model of an FMCW-based UWB chipless RFID reader. The tag is the sole scatterer based on the REP principle and represented by its transfer function. The delay introduced by the cables and the devices is represented by the two transport delay blocks. The transfer function of the tag is the tag mode response of a resonator at a frequency of 4 GHz with a quality factor of 100*

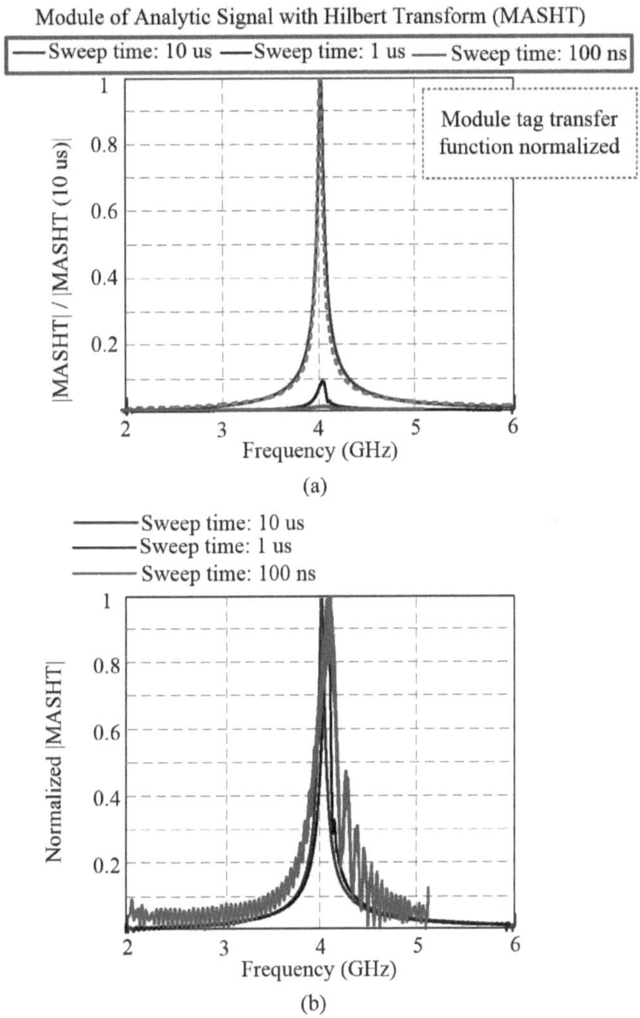


**Figure 2.6.** *Simulation results of the model shown in Figure 2.5 for sweep times of 100 ns, 1 µs and 10 µs. a) Module of the analytical receiving signal obtained with the Hilbert transform. All the curves were normalized with respect to the most favorable sweep time of 10 µs. The green dotted line represents the normalized module of the tag's transfer function. b) The three normalized cases of sweep time of (a). For a color version of this figure, see www.iste.co.uk/garbati/chipless.zip*

The simulation results of the model, depicted in Figure 2.5, are shown in Figure 2.6 for sweep times of 100 ns, 1 µs and 10 µs. The modules of the analytical receiving signal obtained with the Hilbert transform are shown in

Figure 2.6(a). All the curves were normalized with respect to the most favorable sweep time of 10 µs. The green dotted line represents the normalized module of the tag's transfer function.

As shown in Figure 2.6(a), the module of the analytical receiving signal obtained with the Hilbert transform perfectly matches the tag response for a sweep time of 10 µs. At a shorter sweep time, it corresponds to a smaller energized signal [KEY 16a], as shown in Figure 2.6(a). The results of the three normalized cases of sweep time are shown in Figure 2.6(b). From the simulation results, the phase variation $T\theta(f^i)$ introduced by the transfer function is too small to be effectively detected. In equation [2.3], $T\theta(f^i)$ represents a phase variation that corresponds to the high-frequency chirp interrogation signal, and not to the smaller beat frequency $f_p T_{path} BW$. If the tag has the information only coded with tag phase characteristics, then a shorter sweep time is recommended to increase the beat frequency, otherwise a compromise needs to be made.

#### 2.3.1.4. *Conclusion*

From [KOS 12] and the proposed simulation model, a frequency-coded tag can be read with a reader whose architecture is close to an FMCW radar. The sweep time of the transmitting VCO plays a fundamental role in retrieving the tag's transfer function characteristics ($TM(f^i)$, $T\theta(f^i)$). A short sweep time allows for a better implementation of the Hilbert transform and the retrieval of tag phase characteristics $T\theta(f^i)$. However, it shows a lower analytical module signal amplitude, as shown in Figure 2.6(a). A long sweep time produces a small beat frequency $f_p T_{path} BW$; hence, the estimation of $T\theta(f^i)$ is more difficult, but allows for a larger analytical module signal amplitude and thus a better estimation of $TM(f^i)$.

The section will demonstrate how the reading of a time-coded tag using an FMCW approach is easier than the reading of a frequency-coded tag and less error-prone.

### 2.3.2. *Time-coded tag*

#### 2.3.2.1. *Time-coded tag as a collection of steady targets*

A time-based UWB chipless RFID tag can be considered as a different number of steady targets located in the direction of the reader [GIR 13]. This implies the absence of any Doppler effects that may complicate the reading. For the sake of simplicity, a time-coded tag with only one reflector is considered. It may be considered as two ideal targets, the closer responsible for the tag structural mode and the farther responsible for the antenna (tag) mode. If the two targets do not influence each other during a reading process, then equation [2.3] can be independently applied to each of them, yielding equation [2.4] for the tag structural mode and equation [2.5] for the antenna (tag) mode:

$$Y_{ADC}(t) \approx TM_S(f^i)cos\left((2\pi(f_0T_{path-S} + f_pT_{path-S}\mathrm{BW}t\,) - T\theta_S(f^i)\right) \quad [2.4]$$

$$Y_{ADC}(t) \approx TM_T(f^i)cos\left((2\pi(f_0T_{path-T} + f_pT_{path-T}\mathrm{BW}t\,) - T\theta_T(f^i)\right) \quad [2.5]$$

where $TM_S(f^i)$ and $T\theta_S(f^i)$ represent the reflective characteristic of the tag structural mode, and $TM_T(f^i)$ and $T\theta_T(f^i)$ represent the antenna (tag) mode. Since the tag is approximated with two ideal reflective targets, the aforementioned parameters can be considered invariant with frequency. $T_{path-S}$ and $T_{path-T}$ are the total propagation delays between the VCO output and the mixer's RF input, respectively, for the signal reflected back from the structural mode and from the antenna (tag) mode.

In a time-based tag, the information is coded on the timing distances between different targets, and from equations [2.4] to [2.5], it can be retrieved to measure the difference between the two beat frequencies $f_pT_{path-S}\mathrm{BW}$ for the structural mode and $f_pT_{path-T}\mathrm{BW}$ for the antenna (tag) mode. The difference between $T_{path-T}$ and $T_{path-S}$ depends on the propagation delay between the two targets. A time-based UWB chipless RFID tag is assumed to have a resolution in the order of nanoseconds, which results in a frequency difference of:

$$f_{diff} = f_p\mathrm{BW}1ns \quad [2.6]$$

From equation [2.6], the difference between the two beat frequencies linearly depends on the bandwidth of the reader BW and on the sweep time with $\mathrm{f}_p$. A fast FMCW reader with a high bandwidth will read the time-based tag with a better timing resolution, increasing the tag bit capacity. The reader proposed in [KOS 12] and analyzed in this section can be used to read time-based tags, where the Hilbert transform is substituted with a spectral analysis of the ADC output to retrieve the distances between all the generated beat frequencies.

### 2.3.2.2. *Low-cost solution*

The lack of interest in retrieving the transfer function of a time-based tag with $TM(f^i)$ and $T\theta(f^i)$ opens the way for designing a low-cost reader in order to reduce the bandwidth of interest BW. This principle was used in [POP 16b], where a low-cost FMCW reader was proposed for reading time-based tags. The block schematic of the reader and a photograph of its RF front-end are shown in Figure 2.7. The transmitting signal is generated by a VCO whose frequency is stabilized by a fractional synthesizer PLL. Owing to the high power output of the VCO (10 dBm), and the gain of the transmitting and receiving antennas (20 dBi), the reader does not require an output power amplifier. The demodulating block is composed of a mixer and an LPF, with a coupler directly placed on the layout. The tag, shown in Figure 2.7, has been optimized to work in co-polarization, where the reading is performed in an anechoic chamber at a distance of 50 cm. The hardware implementation cost of the reader without the antennas was about 250 € in 2016.

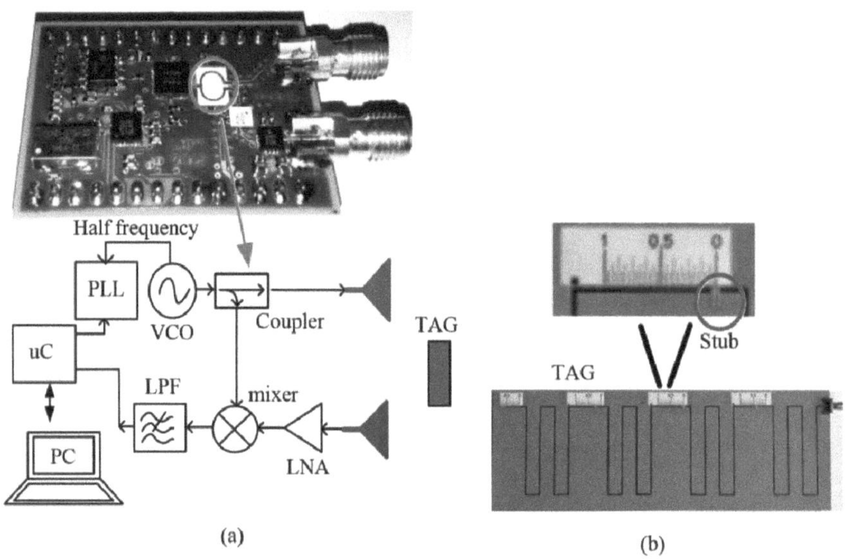


**Figure 2.7.** *a) Photograph of the analog board of a UWB chipless RFID reader proposed in [POP 16b] and its architecture. It is similar to that proposed in [KOS 12]. Its lower cost is due to its limited bandwidth. b) The time-coded tag is composed of four scatterers with stub loads. For a color version of this figure, see www.iste.co.uk/garbati/chipless.zip*

The tag was fabricated on a Rogers RT/duroid 6006 substrate and contains four scatterers with parallel stub loads. The dimensions of the tag are important, which are about 23.7 cm × 8.2 cm. The measurement process described in [POP 16b] uses a calibration tag to remove multiple reflections different from those of the tag reflectors. The tag was successfully tested, but no results have been reported in a practical environment.

The cost of the VCO can be a major obstacle to designing a wideband reader. In [POP 16b], it has a limited bandwidth of 1.25 GHz centered at 7.825 GHz, which indicates an inexpensive solution. The PLL analog devices HMC703LP4E have a built-in frequency sweep mode which is used to control the VCO analog devices HMC508LP5E in a feedback loop. The VCO has two outputs: one is at higher frequency, called the RF output, and the other is at half frequency, which is connected to the PLL to form a loop. In reception after the usual LPF, the microcontroller STM32F407 from STMicroelectronics provides the digital conversion. This solution is solely designed using time-coded tags, as proposed in the literature, and represents the cheapest UWB chipless RFID reader available today.

## 2.4. SFCW-FMCW versus IR-UWB

### 2.4.1. *Introduction*

A UWB chipless RFID reader based on the IR-UWB approach uses a sub-nanosecond UWB pulse to characterize the tag and therefore represents a time-based solution, as shown in Figure 2.1. A schematic of an IR-UWB-based UWB chipless reading system using laboratory equipment is shown in Figure 2.8. A pulse generator triggers the digital signal oscilloscope (DSO) to transmit a sub-nanosecond UWB pulse toward the tag. Then, the DSO measures the backscattered signal from the environment in a bi-static configuration. Alternatively, a diplexer can be used to obtain a monostatic configuration. The IR-UWB approach helps to send an interrogating signal with a maximum instantaneous power compared with the SFCW and FMCW approaches, and interrogates the tag for a shorter period of time.

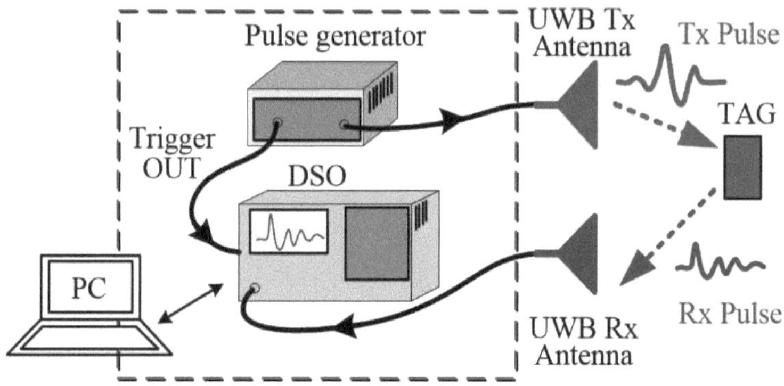


**Figure 2.8.** *A bi-static chipless reading system based on the IR-UWB approach using laboratory equipment. A pulse generator first triggers the DSO to transmit a sub-nanosecond UWB signal toward the tag, and then the DSO measures the backscattered signal from the environment*

The IR-UWB approach shows a shorter reading time due to the absence of a sweep in the frequency domain. The sweep time of a frequency domain solution is directly proportional to the bandwidth of the reader. In [KOS 12], the sweep time was equal to 500 ms. In order to estimate the complete reading time, it is necessary to include all post-processing treatments and an eventual data transfer to a central unit. Moreover, if a sweep-averaging process is used, the reading time will reach several tens of seconds. In [POP 16b], the sweep time was limited to 10 ms; however, its bandwidth was only 1.25 GHz, which means the proposed reader can hardly be used to read high-bit capacity tags (frequency-coded tags).

### 2.4.2. *UWB regulations*

In 2002, the FCC issued its first radio-regulation for the UWB [FCC 17], followed by the ETSI in Europe [ETS 04, ETS 10]. The UWB is an unlicensed bandwidth and therefore for free use. It is characterized by low power transmission where the limitations depend on its applications.

Both the FCC and the ETSI define the limits for the maximum average power spectral density and the maximum peak power. The maximum peak power is defined as a bandwidth of 50 MHz around the maximum measured average spectral density. The latter is measured with a resolution of 1 MHz. FCC and ETSI masks for the maximum peak power and the maximum average power spectral density are shown in Figure 2.9. An IR-UWB reader has a higher instantaneous transmitted power than a frequency domain reader. Thus, the mask for the maximum peak power represents a major constraint for the IR-UWB reader, while the mask for the maximum average power is a major constraint for SFCW and FMCW readers.

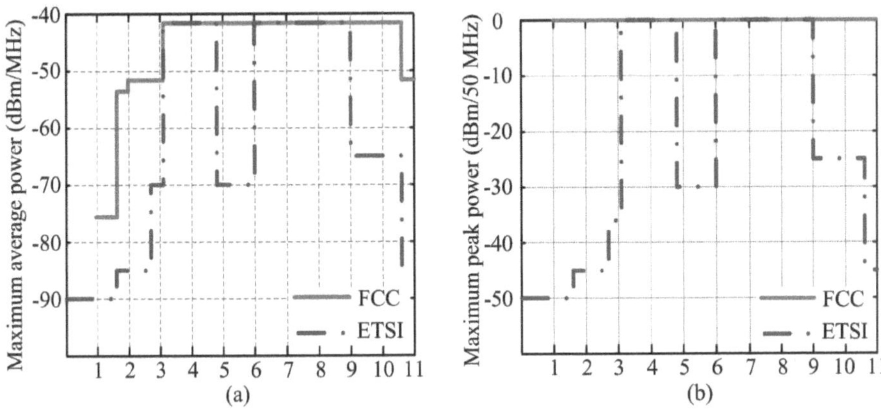


**Figure 2.9.** *FCC and ETSI masks for communication applications with respect to the most favorable case: a) maximum average power and b) maximum peak power. For a color version of this figure, see www.iste.co.uk/garbati/chipless.zip*

### 2.4.3. *Link budget*

A study is proposed for a link budget in the case of a chipless RFID system by taking the regulations into account. The radar equation (Friis) is given by:

$$P_R = \frac{P_T G_T G_R \lambda^2}{(4\pi)^3 R^4} \sigma \qquad [2.7]$$

where $G_T$ is the gain of the transmitting antenna, $G_R$ is the gain of the receiving antenna, $\sigma$ is the radar cross-section of the tag for a given frequency, $\lambda$ is the wavelength of the carrier frequency, $R$ is the distance of the tag, $P_T$ is the reader's transmitting power and $P_R$ is the reader's receiving power.

For a frequency domain reader, the carrier frequency sweeps between a minimum and a maximum value. Thus, the receiving power from the reader also sweeps in frequency at the same period of time, depending on the period of the modulating signal. For the most favorable case, the maximum average power that sweeps in frequency between 3.1 and 10.6 GHz, is −41.3 dBm/MHz, as shown in Figure 2.9(a). It is equivalent to the instantaneous power transmitted by the reader when its sweep time is longer than that of the signal analyzer (SA) which is used to check the compliance with the maximum average power mask. To comply with the regulations, the maximum sweep time for MHz of the SA is 1 ms. If the reader has a sweep time for MHz lower than 1 ms, then its maximum average power can be higher than −41.3 dBm/MHz. Substituting the value −41.3 dBm/MHz in $P_T G_T$ yields equation [2.8] in dBm:

$$P_R = -41.3\ dBm + 10\ log\ 10\ (\frac{G_R \lambda^2}{(4\pi)^3 R^4} \sigma) \qquad [2.8]$$

An IR-UWB reader transmits all the energy of its UWB pulse concentrated in a small fraction of time, while for a frequency domain reader, the energy is distributed in time. As a result, equation [2.7] can be integrated around the band of interest, which is the frequency resolution of the IR-UWB reader (RBW), yielding equation [2.9]:

$$P_R = \frac{G_R \lambda^2 \int_{f_a}^{f_b} P_T G_T df}{(4\pi)^3 R^4} \sigma \qquad [2.9]$$

In equation [2.9], the product $P_T G_T$ must be defined according to the masks. For an IR-UWB approach, the most restrictive mask is the maximum peak power, which, for the worst case, is −34 dBm/MHz according to equation [2.10]:

$$Peack_{Imp}\ (RBW_{New}) = Peack\ (RBW_{50MHz}) + 20\ log_{10}\ (\frac{RBW_{New}}{50\ MHz}) \qquad [2.10]$$

where $Peack(RBW_{50MHz})$ is the maximum peak power, as shown in Figure 2.9 (0 dBm/(50 MHz)), and $20log_{10}(\frac{RBW_{New}}{50\ MHz})$ is the scale factor to express $Peack(RBW_{50MHz})$ at a new resolution of 1 MHz. Equation [2.9] can be expressed in dBm, as shown in equation [2.11]:

$$P_R(dBm) = -34\ \text{dBm} + 10log10(f_b - f_a\ (MHz)) + 10log10(\frac{G_R \lambda^2}{(4\pi)^3 R^4} \sigma) \qquad [2.11]$$

Compared with the results of equations [2.8] (SFCW-FMCW) and [2.11] (IR-UWB), the IR-UWB approach appears to be more favorable, especially for time-coded tags whose spectral characteristics are not of concern, and then $f_b - f_a$ can be as large as possible. For example, if a frequency-coded tag is measured with a frequency resolution of 50 MHz, then equation [2.11] (IR-UWB) becomes:

$$P_R = -17 \text{ dBm} + 10log10\left(\frac{G_R\lambda^2}{(4\pi)^3R^4}\sigma\right) \qquad [2.12]$$

Compared with equations [2.8] and [2.12], the IR-UWB approach shows a 24 dB higher receiving power. To increase $P_R$, the reader should have a short sweep time, which is the case for an FMCW reader. This will decrease the maximum measured average power spectral density over a specified interval of time for a resolution of 1 MHz (1 ms for the FCC and 1 µs–1 ms for the ETSI). However, an FMCW approach shows difficulty in reading a frequency-coded tag, as demonstrated in section 2.3. The transmitting power to the tag can be increased by other power compression methods, such as pulse frequency modulation (PFM), which has not yet been proposed for a chipless RFID reader.

Compared with an IR-UWB reader, a frequency domain reader has a lower input noise bandwidth due to the presence of the IF filter, resulting in a higher sensitivity that may compensate for the lower receiving $P_R$. An IR-UWB reader has a higher noise bandwidth due to the presence of the wideband receiver, which is translated into high sampling noise.

Numerous studies in the literature have exploited the advantages of IR-UWB for chipless applications. These studies have used commercial IR-UWB radars to read time-coded tags. In [RAM 16a], a Time Domain PulsON P400 MRM was used to read a time-coded UWB chipless RFID tag with a read range of 5 m in a practical environment. For all the frequency domain readers proposed in the literature, the read range was less than 1 m in an anechoic environment. The only reader tested in a practical environment was [KOS 12] at a distance of 10 mm. In [RAM 11], the Geozondas GZ6EVK IR-UWB radar evaluation kit, comprising the GZ6E sampler converter and the impulse generator GZ1120ME-50EV, was used to read a time-coded tag at a distance of 50 cm in a practical environment. In [RAM 12b], a Novelda NVA610 evaluation kit for IR-UWB radar applications was used in a bi-static configuration with two tapered-slot Vivaldi antennas, in order to read a time-coded tag at a distance of 130 cm in a practical environment. As analytically demonstrated, an IR-UWB reader is preferable to a frequency domain reader when the UWB regulations are taken into account. It shows an increased power in reception and a reduced reading time. If the UWB regulations are not taken into account, the study of a link budget will no longer be of much significance.

### 2.4.4. *Conclusion*

An FMCW solution with a reduced sweep time is found to be difficult not only with respect to hardware, because of the UWB VCO's limited speed, but also in terms of tag characteristics. Indeed, a frequency-coded tag implemented with the REP technique is no more than a bunch of parallel passive filters. If the sweep in frequency is too fast, these filters wil not have the time to charge themselves and reach the stable state, and the response will be lower in amplitude, as demonstrated in [KEY 16a]. An SFCW reader can be used to read both time- and frequency-coded tags where the reduced $P_R$ can be compensated by a higher sensitivity compared with an IR-UWB reader. In contrast, the long sweep time will translate into high reading time, especially in the presence of sweep averaging [KOS 12]. An IR-UWB architecture is preferable to design a UWB chipless RFID reader that is compliant with international regulations and capable of reading both time- and frequency-coded tags. It has a higher Pr and a reduced reading time due to the absence of any frequency sweeping in transmission, which makes it ideal for a real-time application. The hardware design of an IR-UWB reader has to be optimized in order to reduce its sampling noise (high input noise bandwidth).

## 2.5. Conclusion

In this chapter, the state of the art of UWB chipless RFID readers has been reviewed. The totalities of the proposed solution are based on either the SFCW [PRE 10b] or FMCW [KOS 12, POP 16b] approach, and their main characteristics are outlined in Table 2.1. Both the frequency domain approaches are able to read time- and frequency-coded tags. The SFCW solution provides better reading for frequency-coded tags, while the FMCW solution is ideal for time-coded tags and a low-cost reader design. The tag reading time exceeds 1 s depending on the VCO sweep time and averaging.

| | SFCW | FMCW | |
|---|---|---|---|
| | **[PRE 10b]** | **[KOS 12]** | **[POP 16b]** |
| **VCO sweep time** | No info | 500 ms | 10 ms |
| **Bandwidth (GHz)** | 5–9 | 4–8 | 1.25 |
| **Estimated component cost (€)** | 4,500 | 650 | 250 |

**Table 2.1.** *Main characteristics of the major UWB chipless RFID readers proposed in the literature, which are based on the frequency domain approach. The cost estimation does not consider the antennas used for the given measurement results*

The IR-UWB architecture has been introduced to provide a faster reading because there is no sweep frequency process in transmission. Based on the international UWB regulations, a link budget has been proposed, which takes both frequency domain and IR-UWB reader approaches into account, with IR-UWB showing a higher receiving power. However, its hardware architecture has to be optimized in order to reduce the sampling noise.

The design of a first reader version using the IR-UWB approach will be discussed in the following chapter.

3

# IR-UWB Chipless RFID Reader Design

## 3.1. Introduction

This chapter presents the design of the first version of an IR-UWB chipless RFID reader. It is fully functional and the first custom-published chipless RFID reader based on the IR-UWB approach. This chapter is organized as follows:

– Section 3.2 introduces an IR-UWB reader system based on test equipment to read frequency-coded UWB chipless RFID tags. It is a real-time system with a realization cost and overall dimensions that are not suitable for a commercial solution.

– Section 3.3 introduces the sequential equivalent time principle for designing low-cost electronic systems. Thus, all of the proposed designed readers in this book are based on the sequential equivalent time principle.

– Section 3.4 presents an intermediate fully working reader version.

– Section 3.5 describes the hardware architecture of the integrated reader version, which was mainly designed by the first author of this book. It shows a substantial reduction in the retail costs of its hardware components, and better performance. The reader's characterization emphasizes a possible improvement to the design of a more high-performing reader.

– Section 3.6 presents some measurement results of frequency-coded tags with the reader introduced in section 3.5.

– Section 3.7 draws the conclusion of this chapter.

## 3.2. IR-UWB reading system based on test equipment

The IR-UWB approach was introduced in Chapter 2. The transmitting signal is a sub-nanosecond UWB pulse that can be used to read both time- and frequency-coded tags.

The reading system of a UWB chipless RFID tag is shown in Figure 3.1(a). The reader is composed of a pulse generator, a device to measure the tag response in the time domain, such as a DSO, and a PC that can be used as a control and post-processing unit. A low-noise amplifier (LNA) can be connected at the input of the DSO. The pulse generator triggers the DSO during the acquisition process. The corresponding test bench used to read chipless tags is shown in Figure 3.1(b). The DSO used is Agilent 912004A, with an analog bandwidth of 12 GHz and a sampling rate of 40 GSa/s. The pulse generator is Picosecond 3500D, which generates pulses with 65 ps of full width at half-maximum (FWHM) and an amplitude of 8 V on a 50 Ω load. The tag is the depolarizing REP tag introduced in [VEN 13a] and presented in Chapter 1. The tag is placed at a distance of 15 cm from the end (the closer point to the tag) of the dual-access Satimo QH2000 antenna. Measurements were performed in an anechoic chamber to reduce external EM (electromagnetic) disturbances as much as possible. The PC is connected to the UWB slave port of the DSO, and a graphical user interface (GUI) is developed, which is described in Appendix 1, to control the DSO.

The tag is read in the cross-polarization configuration, as shown in Figure 3.1(a), using the background subtraction technique introduced in Chapter 1. The latter consists of two measurements: one with the tag (tag measurement) and the other without the tag (background measurement). After processing, a subtraction between the two results is performed in the time domain to isolate the tag contribution from the measured signals. The tag measurement and the subtraction result are shown in Figure 3.2(a) and (b), respectively. The tag contribution is a less-powerful part of the signal, while the most significant part is represented by the direct coupling between the antenna ports (even though they are characterized by an isolation of about 40 dB). This represents the leakage of the presenting reading system, as shown in Figure 3.1(a) by a green dashed line. To retrieve the tag ID, it is sufficient to analyze the signal shown in Figure 3.2(b) in the frequency domain and the correct frequency resonant positions (peaks in the spectrum) shown in Figure 3.2(c).

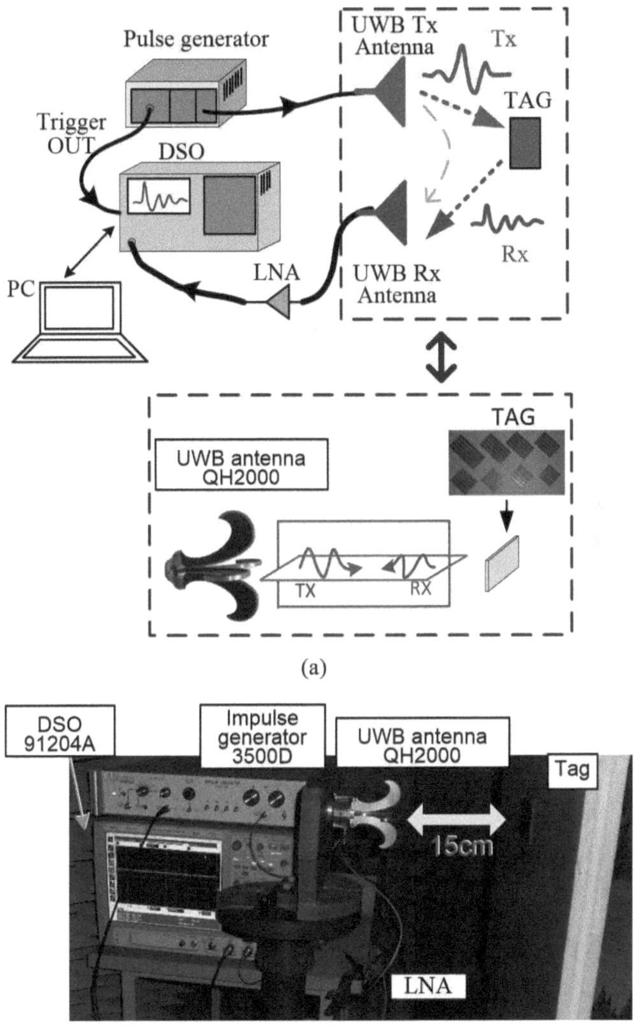


**Figure 3.1.** *A bi-static chipless reading system based on the IR-UWB approach and composed of test equipment. a) Schematic of the actual reading system used. The reading was performed in the cross-polarization configuration using the dual-access Satimo QH2000 antenna. b) A photograph of the test bench in an anechoic chamber. The tag is the depolarizing REP tag introduced in [VEN 13a] and presented in Chapter 1. It is placed at a distance of 15 cm from the antenna. For a color version of this figure, see www.iste.co.uk/garbati/chipless.zip*

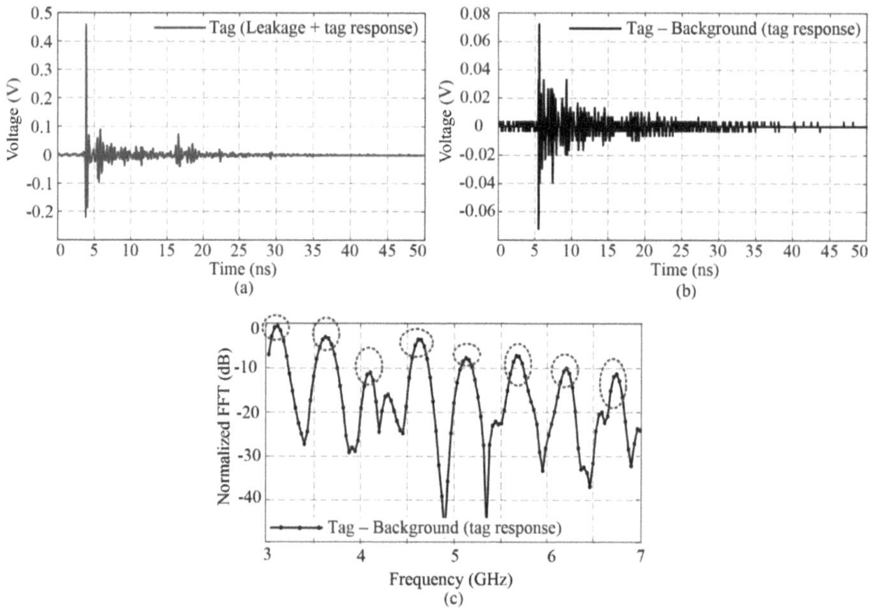


**Figure 3.2.** *Measurement results in the time domain for the test bench presented in Figure 3.1. a) Tag measurement, where the tag contribution and antenna leakage are part of the signal. b) Background subtraction measurement to isolate the tag contribution from antenna leakage. c) Frequency analysis of the tag contribution; the red dotted circles represent the position of the eight tag resonant frequencies (ID)*

## 3.3. Sequential equivalent time principle

The reading system, shown in Figure 3.1, is a real-time implementation, thanks to the DSO's maximum sampling rate of 40 GSa/s. To design a reader at low cost, an equivalent time reading approach is used. In the sequential equivalent time approach, the reader transmits the same interrogating signal toward the tag at intervals corresponding to the pulse repetition frequency (PRF) period. Subsequently, the reader samples only a few points of each backscattered signal from the tag to reconstruct the entire signal. This is possible as long as the reading scenario is stationary between the different interrogations. It should also be noted that a measurement is composed of many contributions, such as the response from the tag, the environmental contribution and the coupling between the antennas (leakage). This procedure is performed with a much lower sampling frequency than that required in a real-time approach, which follows the Nyquist–Shannon sampling theorem.

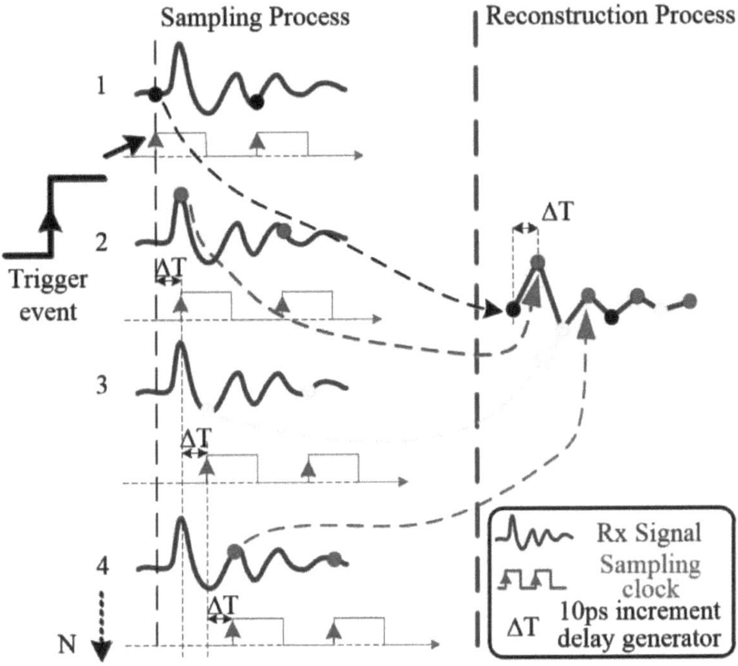


**Figure 3.3.** *Sequential equivalent time principle. The signal of interest (Rx impulse) is sampled with a low sampling frequency (sampling clock), which does not satisfy the Nyquist–Shannon theorem. Therefore, only a few points of the tag response are acquired. Different pulses are transmitted to the tag, and for each tag response, the sampling clock is shifted within a few picoseconds, ΔT. The synchronization is established based on a trigger event. For a color version of this figure, see www.iste.co.uk/garbati/chipless.zip*

The sequential equivalent time principle is summarized in Figure 3.3, where the left-hand side shows the sampling process and the right-hand side the subsequent tag response reconstruction. Between the distinct acquisition processes, the sampling clock is shifted within a few picoseconds, ΔT, to sample all of the points corresponding to the measured signal. Each acquisition process starts after a trigger event, which is usually generated by the reader control unit or the pulse generator. Finally, in the reconstruction process, all of the samples in the reader's memory are rearranged together to retrieve the measured signal. The equivalent sampling frequency of the system is therefore 1/ΔT. With this approach, the reader can emulate a high-speed sampling system, as shown in Figure 3.1. The reading procedure can be executed with an averaging by repeating the tag's interrogation sequence. It is useful to improve the signal-to-noise ratio (SNR) and to avoid external disturbances from other EM signals [VEN 14]. In the following section, the intermediate fully working IR-UWB reader version is presented.

## 3.4. Intermediate reader version

Before the first author of this book joined the LCIS (Laboratoire de Conception et d'Intégration des Systèmes, France) laboratory in February 2014, the research group of the second author had already developed a first fully working reader version during the two years of a project called Gravit, started in December 2010. This section first details the schematic of the prototype and then relates it to Figure 3.3 in order to explain the hardware implementation of the sequential equivalent time principle.

### 3.4.1. *Schematic*

The retail cost of evaluation boards and hardware components that constituted the reader prototype was about 5,638 € after tax (see Table 3.1), without considering the antennas. The block diagram of the reader is shown in Figure 3.4. The hardware of the reader was developed for implementation in a bi-static configuration in order to reduce the leakage between the transmitting and receiving hardware parts. The pulse generator is triggered by the digital section, whose output is directly connected to the transmitting UWB antenna. In reception, the RF front-end is made up of an LNA in series with a sample-and-hold amplifier, henceforth referred to as the S/H amplifier, and an analog-to-digital converter (ADC), whose output is connected to the digital section.

A photograph of the reader is shown in Figure 3.5. The LNA is the Miteq AMF-4D-00101200-34-13P, which has a gain of 40 dB and a bandwidth of 0.1–12 GHz. The maximum noise level is 3.4 dB and the minimum output 1 dB compression point is 13 dBm. The required supply voltage is 15 V, with a current consumption of 325 mA. The S/H amplifier is the Inphi 1321TH integrated on its evaluation board. This component has an input bandwidth of 0–13 GHz and supports a maximum sampling rate of 2 GSa/s. The differential output of the S/H amplifier is connected to the analog input of a home-designed four-layer board on an FR-4 substrate which is part of the reader's RF front-end. The main components of the reader are the ADC and the delay generator.

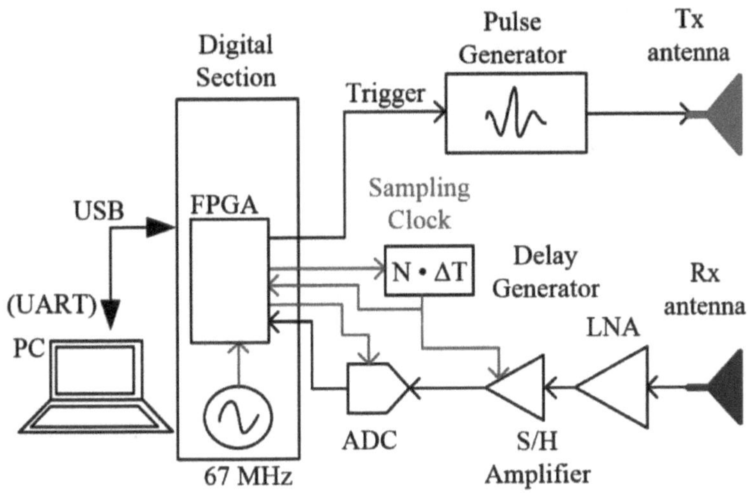


**Figure 3.4.** *Architecture of a fully working designed reader during the project Gravit. It has a bi-static configuration and is composed of three parts: a UWB pulse generator, a digital section and the receiving RF front-end. For a color version of this figure, see www.iste.co.uk/garbati/chipless.zip*

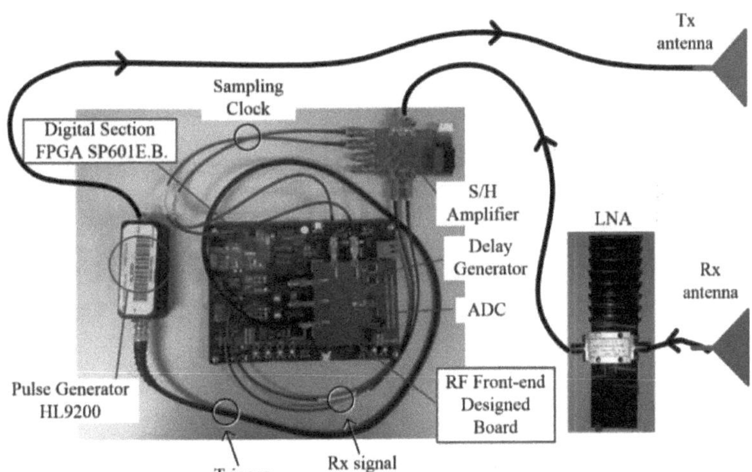


**Figure 3.5.** *A photograph of a fully working designed reader during the project Gravit. The pulse generator is the HL9200 from Hyperlabs. The digital section is enclosed in a Xilinx SP601 evaluation board for the FPGA Spartan 6 XC6SLX16. The four-layer RF front-end board was fabricated on an FR-4 substrate. The LNA is the Miteq AMF-4D-00101200-34-13P. The S/H amplifier is the Inphi 1321TH integrated on its evaluation board. For a color version of this figure, see www.iste.co.uk/garbati/chipless.zip*

The ADC is the low-cost AD9215 from analog devices, which has a vertical resolution of 10 bits. The ADC with its differential inputs can be either 1 or 2 $V_{P-P}$. It has an input bandwidth of 300 MHz and a maximum sampling rate of 80 MSa/s. The delay generator is implemented with two chips (MC100EP196 from ON Semiconductor) that are connected in series, each of which has an input digital bus of 10 lines to generate internal delays. These chips are able to delay a 50% duty-cycle clock signal between 2.4 and 12.4 ns, with a minimum time step (ΔT) of 10 ps (see Figure 3.3). The sampling clock period of the reader is of about 15 ns (67 MHz); therefore, the use of the two chips is indispensable.

The digital section is enclosed in a Xilinx SP601 evaluation board for an FPGA Spartan 6 XC6SLX16. The pulse generator is the HL9200 from Hyperlabs, which can generate a wide-spectrum pulse in the frequency domain that can be extended into the UWB. The generated pulse has a rise and fall time of 50 ps, a pulse width of 70 ps and an amplitude of 2 V on a 50 Ω load with a random jitter of 3 ps root mean square (RMS). The required supply voltage is 9 V, with a current consumption of 200 mA.

### 3.4.2. *Implementation of the sequential equivalent time principle*

The aim of the reader was to measure UWB chipless RFID tags, according to the sequential equivalent time principle summarized in Figure 3.3. The FPGA is the controller of the acquisition process, which triggers the pulse generator, sets up the value of the delay generator and memorizes the ADC output. It is controlled by a PC with a Matlab® program in the Windows environment, which selects the averaging and retrieves the measured signal in the time domain. The delay generator's resolution of ΔT = 10 ps corresponds to an equivalent sampling rate of 100 GSa/s (1/ΔT). The S/H amplifier is able to sample a UWB input signal and output a constant signal for an entire sampling clock period. It sends a signal to the ADC to convert the S/H amplifier output into digital signals. The trigger event of the pulse generator is synchronized with the sampling clock of the S/H amplifier at the input of the delay generator. This means that during the acquisition process, the internal delay of the delay generator varies with a time step of ΔT, according to the sequential equivalent time principle summarized in Figure 3.3. Indeed, the delay between the trigger event of the pulse generator and the sampling clock of the S/H amplifier varies with a time step of ΔT.

The application software written in Matlab® is based on different scripts generated by the GUI. It expects the execution of a tag measurement according to the background subtraction technique. The GUI is designed to read the frequency-coded tag, presented in [VEN 13a], based on coupled short dipoles using the REP and cross-polarization approaches, as shown in Figure 3.1. The tag was characterized in section 1.3 of Chapter 1 and shown in Figure 1.10. The Matlab® program uses a fast Fourier transform (FFT) approach to retrieve the tag ID. The Matlab® GUI is described in greater detail in Appendix 2.

The first fully working reader version demonstrated the feasibility of the project, without taking optimization into account. The reader was able to read the tag with the same Satimo QH2000 antenna, and could decode the tag information up to 5 GHz which corresponds to the first four resonant frequencies over eight. Table 3.1 presents the main characteristics of the reader.

| **Reader's retail cost after tax** | |
|---|---|
| **Component** | **Retail cost after tax (2013)** |
| LNA (Miteq AMF-4D-00101200-34-13P) | 2,619 € |
| S/H amplifier (Inphi 1321TH E.B.) | 1,469 € |
| Realized board components | ≈200 € |
| Digital board (Xilinx SP601 E.B.) | 350 € |
| Pulse generator (Hyperlabs HL9200) | 1,000 € |
| **Total cost** | 5,638 € |
| **Reading performance** | |
| **PRF** | 1.04 MHz |
| Input bandwidth | 0.1–12 GHz |
| Acquisition frame duration | 120 ns |
| Frame position | Fixed |
| Equivalent sampling period (ΔT) | 10 ps |
| Averaging | 1–128 |
| Reading time | ≈12.3 s + Averaging factor × 0.5 s |
| S/H amplifier's sampling clock period jitter | 13.22 ps RMS (115.67 ps p-p) over 1,000 cycles |

**Table 3.1.** *Components' retail cost after tax of a fully working designed reader during the project Gravit (without considering the antenna), and performance specifications*

The most expensive component of the reader is the LNA with its cost equal to 2,619 €, followed by the S/H amplifier and the pulse generator. The input bandwidth of the reader is in the range of 0.1 and 12 GHz, and the PRF is 1.04 MHz. The acquisition frame position is fixed, and the equivalent sampling period is $\Delta T = 10$ ps. The reading time is 12.3 s plus 0.5 s multiplied by an averaging factor of 128.

From Figure 3.3, it can be seen that if the sampling clock is affected by a random jitter, then the acquisition process will be affected by sampling noise. As the power of the generated sampling noise increases with the reading signal frequency [GEN 08], its reduction is essential in a UWB chipless RFID reader. From Table 3.1, it can be seen that the S/H amplifier sampling clock of the reader has a random period jitter of 13.22 ps as standard deviation (RMS) and a maximum peak-to-peak jitter of 115.67 ps over 1,000 cycles. The integrated reader version is described in the following sections.

## 3.5. Integrated reader design

This section presents the integrated reader version, which consists of three main parts: an RF front-end board, the Xilinx SP601 evaluation board for the digital section and the pulse generator for the emission part. The reader has an identical frame length, an equivalent sampling period ($\Delta T$) and a maximum averaging factor as the previous version (see Table 3.1). The main improvements made are the reduction in the reader's realization cost and the improvement in the reading acquisition. The former was obtained with the integration of a low-cost LNA and the S/H amplifier on a unique RF front-end board. The latter was obtained with a significant improvement to the S/H amplifier and ADC connections, and with a substantial reduction of the sampling clock random period jitter.

### 3.5.1. *Hardware architecture*

The block schematic of the reader, shown in Figure 3.6, is invariant with respect to the previous reader version (see Figure 3.4). The sampling clock of the reader is 67 MHz with the PRF of about 1 MHz. When comparing Figure 3.5 with Figure 3.6, the new version is a more integrated solution, which aims to reduce the reader's realization cost and the overall reader dimensions.

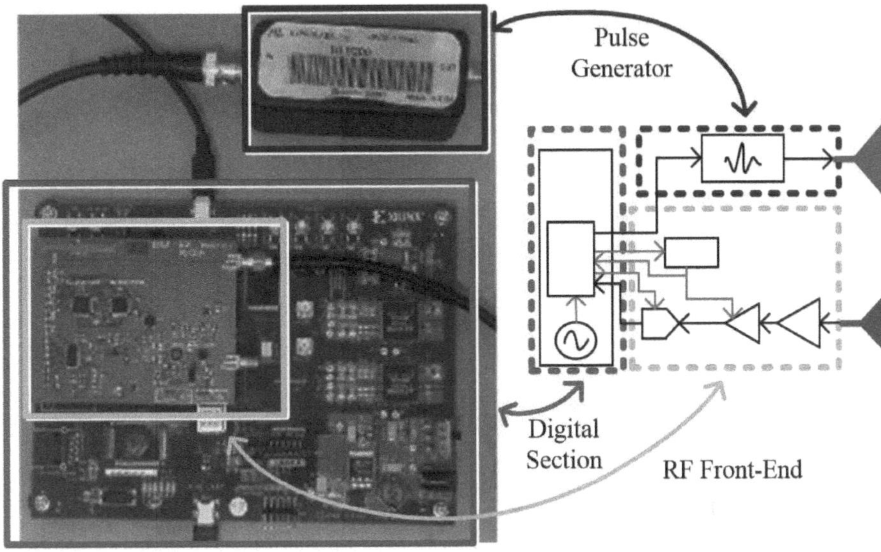


**Figure 3.6.** *A photograph of the reader. The pulse generator is the HL9200 from Hyperlabs. The digital section is enclosed in a Xilinx SP601 evaluation board for the FPGA Spartan 6 XC6SLX16. The four-layer RF front-end board is fabricated on an FR-4 substrate. For a color version of this figure, see www.iste.co.uk/garbati/chipless.zip*

The more high-performing LNA Miteq AMF-4D-00101200-34-13P is substituted with the HMC754LP4E from analog devices. It operates in the bandwidth of 1–11 GHz, and has a gain of 16.5 dB at 6 GHz and 14 dB at 4 GHz. The noise figure is 1.5 dB at 4 GHz, and the output compression point is 18 dBm at 1–6 GHz. The AMF-4D-00101200-34-13P costs about 2,619 € (2013), whereas the cost of the HMC754LP4E is within 100 €. The components' integration work was carefully executed to improve the reader's SNR as much as possible. This involved the use of different components, such as the driver AD4930, the optimization of the RF front-end schematic and layout, and a better implementation of the reader's FPGA architecture.

The RF reader front-end board has two SMA connectors, one for triggering the pulse generator and the other as input for the external UWB-receiving antenna. A photograph of the RF front-end is shown in detail in Figure 3.7. The four-layer board is fabricated on an FR-4 substrate, which measures 88 × 70 mm (L × H).

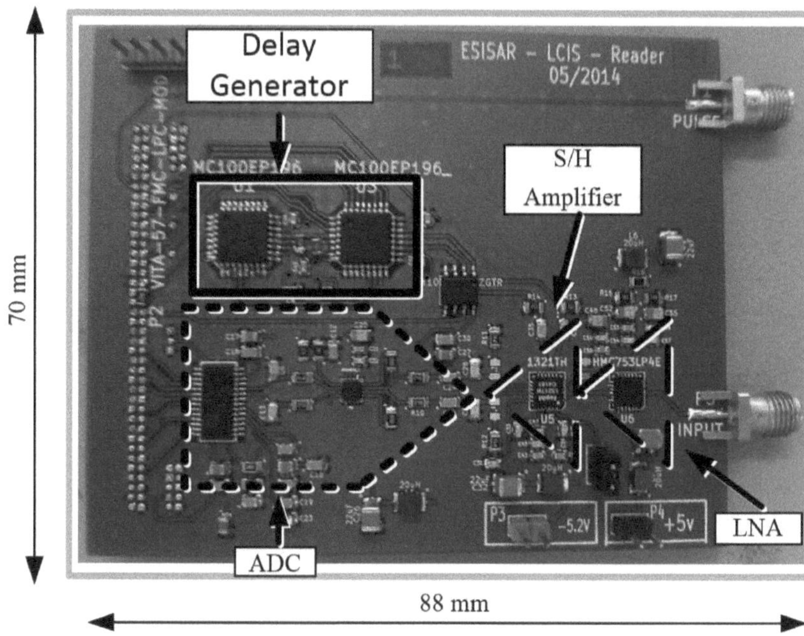


**Figure 3.7.** *Photograph of a realized RF front-end of the reader shown in Figure 3.6. It is a four-layer board fabricated on an FR4 substrate, which measures 88 × 70 mm (L × H). For a color version of this figure, see www.iste.co.uk/garbati/chipless.zip*

According to the previous reader version, the delay generator was implemented with two MC100EP196 chips connected in series. The input of the first chip is the sampling clock to be delayed (see Figure 3.4), which is directly sourced from the FPGA. For both chips, the digital control bus, for setting up the delay value, is connected to the FPGA. The output sampling clock signal of the second chip is directly connected to the sampling clock input of the S/H amplifier and with the FPGA (see Figure 3.4). To improve signal common noise rejection, the sampling clock connection between the FPGA and the delay generator, and between the delay generator and the S/H amplifier, are made up of balanced signals based on the digital standard positive emitter coupled logic (PECL). A PECL–TTL converter was used to trace the sampling clock at the output of the delay generator inside the FPGA.

The S/H amplifier input can be a UWB signal, which is sampled with a sampling clock of only 67 MHz. The S/H amplifier varies its output only at the sampling clock event that occurs every 15 ns (about 1/67 MHz). It gives the time at the following ADC to convert this low frequency signal, as shown in Figure 3.8. The ADC is the low-cost AD9215 from analog devices, which is also used in the previous version, with a vertical resolution of 10 bits. The LNA has a single input/output mode, while the input of the sampler can be driven in both single and differential modes. No balun was used to convert the LNA output into a differential one. In fact, it is difficult to find a commercially available balun to cover all of the UWB in the range of 3.1 to 10.6 GHz that is cheap and small in size.

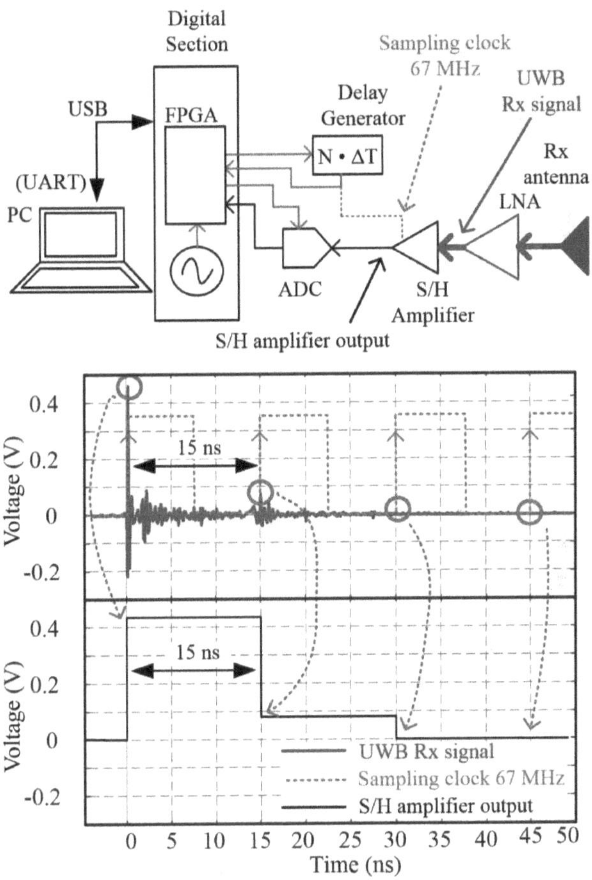


**Figure 3.8.** *S/H amplifier's principle of operation. A UWB input signal (UWB Rx signal) is sampled with a low frequency clock signal (sampling clock 67 MHz). Thus, its output (S/H amplifier output) is a lower frequency content signal. For a color version of this figure, see www.iste.co.uk/garbati/chipless.zip*

### 3.5.2. *Improvement of S/H amplifier–ADC connection*

The proper interconnection between the S/H amplifier and the ADC input is the subject of interest. The schematic of this interconnection of the previous reader version, presented in Figure 3.5, is shown in Figure 3.9. The differential output of the S/H amplifier has an equivalent impedance of 100 Ω and a dynamic of 1 $V_{P-P}$. The input impedance of the ADC has a differential resistance of 7 kΩ, and each pin has a capacitance of 2 pF. The ADC input requires a common voltage of about 1.5 V to operate at maximum performance levels, which represents half of its analog supply voltage. Thus, the output of the sampler and the input of the ADC must not be directly connected.

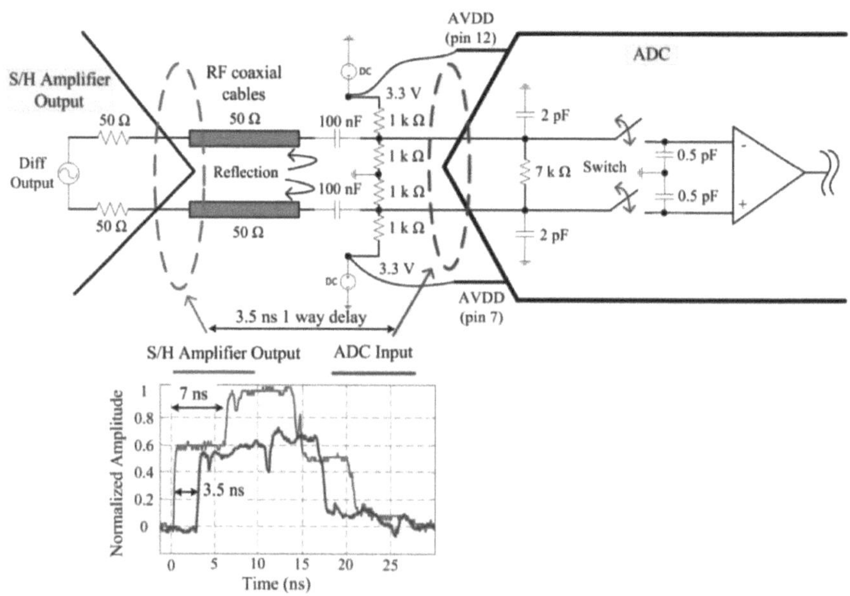


**Figure 3.9.** *Schematic of the S/H amplifier and ADC interconnection of the reader version presented in Figure 3.5. The mismatching at the ADC input pins creates important reflections toward the S/H amplifier output, as shown with the plotted curves. As the S/H amplifier and the ADC are in two different boards, RF coaxial cables are used for their interconnections. The delay between the S/H amplifier output and the ADC input is 3.5 ns. The reader sampling clock period is 15 ns. For a color version of this figure, see www.iste.co.uk/garbati/chipless.zip*

As shown in Figure 3.5, the S/H amplifier and the ADC are located in two distinct boards; therefore, RF coaxial cables are used for their interconnections. The sampler and the ADC are AC coupled with two capacitors of 100 nF, and the common voltage of the ADC input is settled with a voltage divider using two resistances of 1 kΩ. This solution is straightforward but has some problems. The voltage divider directly connects the analog supply pins (AVDD pin 12/7) to the ADC analog input. It may represent a low impedance path toward the input of the ADC for the supply noise. The ADC analog input impedance varies during the acquisition process, depending on its state (sampling or holding state). In a sampling state, its analog input impedance decreases because the two switches (see Figure 3.9) become closed. It will force the output of the sampler to drive the analog input of the ADC. Finally, the output of the sampler is not closed to a proper differential resistance of 100 Ω, creating an impedance mismatching. Its negative effects are shown in Figure 3.9 with the measurement of the sampler output on its evaluation board and the input of the ADC. As shown in Figure 3.8, the sampler output is expected to be a step signal with an interval of about 15 ns, which is equal to the reader sampling period (1/67 MHz). Instead, in Figure 3.9, the signal does not appear to be constant with a higher interval of about 22 ns. This is caused by the reflected signal, at the aforementioned mismatching, where the two-way propagation delay is about 7 ns (RF coaxial cables, PCB layout and components). It can be defined as the sum of two equal signals of duration 15 ns (sampling clock), with a delay of 7 ns.

In the new reader version, the S/H amplifier and the ADC are on the same RF front-end board (Figure 3.7). To improve the performance of their interconnection, a driver AD4930 from analog devices is used. It is a differential amplifier designed to be interfaced with the ADC from the same maker. The simplified schematic of the implemented solution is shown in Figure 3.10. The driver input pin, denoted by $V_{OCM}$, settles the driver output common mode voltage, and with the external two resistors, denoted by *Rf*, it is possible to configure a differential gain.

The two capacitors of 10 ps at the output of the S/H amplifier pins, and the two at the ADC analog inputs pins, are used to improve the AD4930 robustness against instability. The sampler output is perfectly matched and does not require driving the ADC input in a sampling state. The use of the driver also decreases the effect of supply noise at the ADC input [ANA 17]. An evaluation board is fabricated on an FR-4 substrate to study the driver effects on the signal acquisition, as shown in Figure 3.10, and the dimensions of the board are 110 × 100 mm (L × H). In the following section, the reader's FPGA implementation will be discussed in detail.

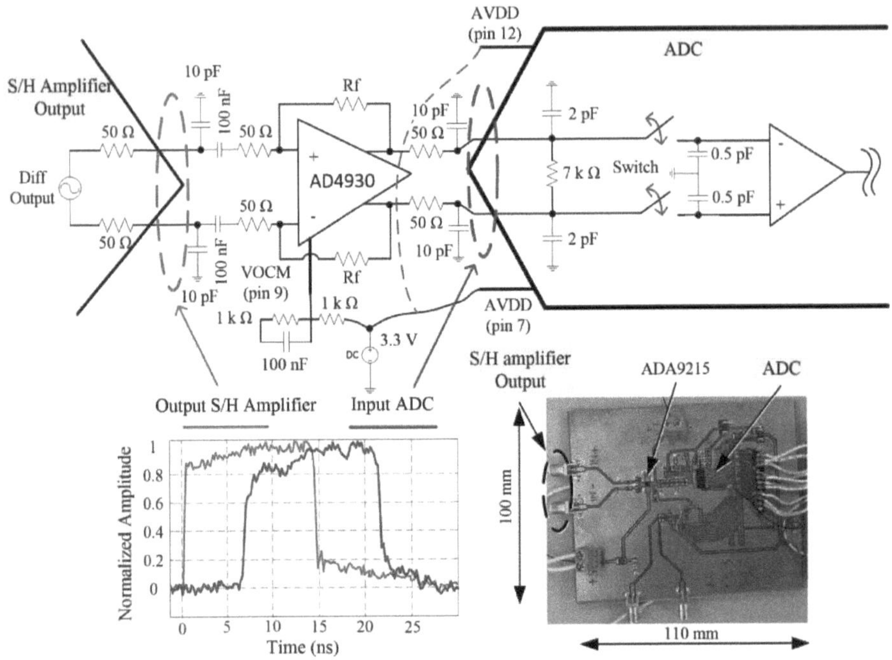


**Figure 3.10.** *Schematic of the S/H amplifier and ADC interconnection in the new reader version. The use of the driver AD4930 helps to reduce impedance mismatching and increases supply noise rejection. An evaluation board is fabricated on an FR-4 substrate. For a color version of this figure, see www.iste.co.uk/garbati/chipless.zip*

### 3.5.3. *FPGA architecture*

The RF front-end is controlled by the digital board SP601 through the FPGA hardware architecture. An FPGA device provides flexibility, and fast dedicated hardware blocks can be created with VHSIC Hardware Description Language (VHDL) scripts. In the proposed reader, the FPGA implements a microprocessor-based embedded system with the soft-core Microblaze processor version 8.20.a. It is connected to different peripherals: some are internal at the chip and others are external, such as Double Data Rate 2 (DDR2) memory, universal asynchronous receiver transmitter (UART), push button and LEDs. Among the internal peripherals, there is a clock synthesizer, an interrupt controller and, the most important, a custom peripheral called Reader_0. The block Reader_0 is described with VHDL and only communicates with the RF front-end. As a result, it is set up by the microprocessor that gives information about the selected averaging. Inside the Reader_0 block, a first-input first-output (FIFO) memory is also present to acquire the output of the ADC.

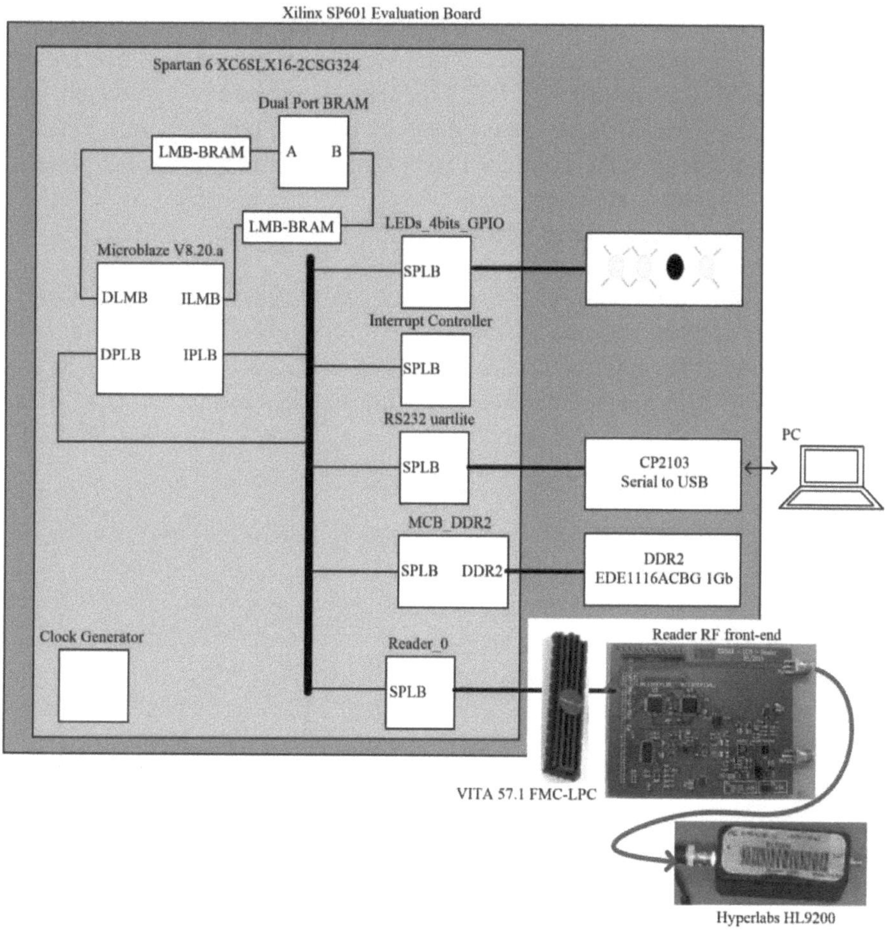


**Figure 3.11.** *Block schematic of an FPGA-embedded system, with the main peripherals of the digital board SP601 used to implement the reader. Only the custom block Reader_0 communicates with the analog board and triggers the pulse generator. For a color version of this figure, see www.iste.co.uk/garbati/chipless.zip*

The block schematic of the FPGA-embedded system, with the main peripherals of the digital board SP601 used to implement the reader, is shown in Figure 3.11. The Xilinx ISE Design Suite 13.2 is used to design, simulate and debag the FPGA architecture. The suite offers many applications that ease the development process. Among them, the ISE Design Tool is the most indicated tool to implement custom hardware blocks. The Platform Studio creates the real microprocessor embedded system, with the addition of all blocks immediately available from Xilinx, and a Software Development Kit for software development and debugging. Inside the FPGA Spartan 6 XC5SLX16, the Microblaze is connected to a local dual-port block random access memory (BRAM) of 8 kb with two fast local memory buses (LMBs), one for data and the other for instructions, realizing a Harvard memory architecture. Two LMB-BRAM interface controllers are needed to connect the Microblaze to the local BRAM.

A Processor Local Bus (PLB) 4.6 is used to connect the microprocessor to the rest of the peripherals, with the exception of the clock generator. The PLB provides a bus control unit, a watchdog timer and a separate address, write and read data path. All of the peripherals connected to the Microblaze are accessed with memory mapped, where the address range for the LMB and the PLB must not overlap.

The LEDs 4-bit GPIO (General Purpose Input/Output) is the interface between the PLB and GPIO to turn four LEDs on and off. They are useful in understanding the status of operation of the reader and also during software debugging. The RS232 Uartlite block is able to connect a PLB bus interface with an external device equipped with UART. In this project, it was used to interface the microprocessor with the external PC. The chip CP2103 from Silicon Labs resides on the SP601 board and is able to interface the Universal Serial Bus (USB) up to version 2.0, with a UART device. All of the application software on the PC was developed for the Microsoft operative systems and work with existing COM port PC applications. The communication with the PC has a Baud Rate (BR) of 460.800 kb/s, with 8 bits for each data frame and without a parity bit. As the Spartan 6 presents limited resources in terms of available memory, the use of an external memory is necessary. The internal BRAM with a capacity of 8 kb is mainly used as the cache memory, while the external DDR2 of 1 Gb is used as the principal memory in which the application software is utilized. The interface MCB_DDR2 can connect the DDR2 memory to the PLB bus. As this solution is not optimal for speed, the PLB is not accounted for a fast bus; however, for this prototype version of the reader, the optimization of the reading time is not a priority.

The Reader_0 is the custom block developed with ISE Design Tool. It directly communicates with the analog board and is also responsible for triggering the pulse generator. It contains a FIFO memory, which is directly connected to the output of the ADC, a main control unit, some blocks for synchronization purposes and a PLB slave interface. The Reader_0 memorizes the ADC output, sets up the internal delay of the delay generator chips and triggers the pulse generator.

The clock generator exploits the Spartan 6 clock resources, as a Phase-Locked Loop (PLL) and a Digital Clock Manager (DCM), available inside the Clock Management Tiles (CMTs) to generate different clock signals for the available interfaces. According to the developed project, the FPGA external chip clock input was the onboard 200 MHz crystal oscillator SiT9102AI-243N25E200.00000 from SiTimes. From the 200 MHz input, using the clock generator block, a 67 MHz clock and two oppositely phased 600 MHz clocks were created. The higher clock rate was used for the external DDR2 memory and the internal controller MCB_DDR2, and the lower one was used for the rest of the components, including the block Reader_0. The interrupt controller is able to concentrate multiple interrupt inputs into one single interrupt output. It is interfaced with a PLB slave interface and presents one output directly connected to the microprocessor interrupt input. An interrupt controller is used to free up the microprocessor from the pulling operation.

### 3.5.4. *Reader specifications*

The costs and performance of the advanced reader are presented in Table 3.2. With respect to the previous version, it shows a halved retail component cost, from 5,638 € to 2,350 €, and a halved sampling clock period jitter. The latter was obtained by working on the FPGA implementation, with proper utilization of internal intellectual property (IP) such as PLL and delay-locked loop (DLL) [XIL 15a], and with I/O dedicated resources [XIL 15b]. Moreover, the connection between the delay generator chips and the S/H amplifier was improved using proper components. Finally, the layout design also played an important role. In addition to all the improvements made to this reader version, some problems with signal synchronization still persist, as with the previous reader version. They are addressed in the following chapter.

| Reader's retail cost after tax | |
|---|---|
| **Component** | **Retail cost after tax (2013*/2014**)** |
| S/H amplifier (Inphi 1321TH E.B.) | 800 €** |
| Realized board components | ≈200 €** |
| Digital board (Xilinx SP601 E.B.) | 350 €* |
| Pulse generator (Hyperlabs HL9200) | 1,000 €* |
| **Total cost** | 2,350 € |
| **Reading performance** | |
| PRF | 1 MHz |
| Input bandwidth | 1–11 GHz |
| Acquisition frame duration | 120 ns |
| Frame position | Fixed |
| Equivalent sampling period (ΔT) | 10 ps |
| Averaging | 1–128 |
| Reading time | ≈12.3 s + Averaging factor × 0.5 s |
| S/H amplifier's sampling clock period jitter | 7.34 ps RMS (67.37 ps p-p) over 10k cycles |

**Table 3.2.** *Components' retail cost after tax of a presenting fully working designed reader (without considering the antenna), and performance specifications*

## 3.6. Measurement results of frequency-coded tags

The reader described herein was tested using the frequency-coded tag introduced in [VEN 13a], as shown in Figure 3.12. The two tags have different resonant frequency positions in the bandwidth of 3–7 GHz. The second tag has the first seven resonant frequencies of 100 MHz, respectively, which are shifted towards a higher frequency (right), while it is 200 MHz for the eighth. The two tags were first measured in turn in an anechoic chamber with the test equipment, as shown in Figure 3.1, and then with the proposed reader. The tags were placed on a foam support at a distance of 15 cm from the antenna. The dual-access dual-polarization Satimo QH2000 antenna and the dual measurement procedure (background subtraction) were used, as described in section 3.2. The reference measurement system (test equipment) used the pulse generator HL9200 (the same as in the proposed reader in transmission). The DSO Agilent 91204A is used in reception, as shown in Figure 3.1, and the measurement bench is shown in Figure 3.12.

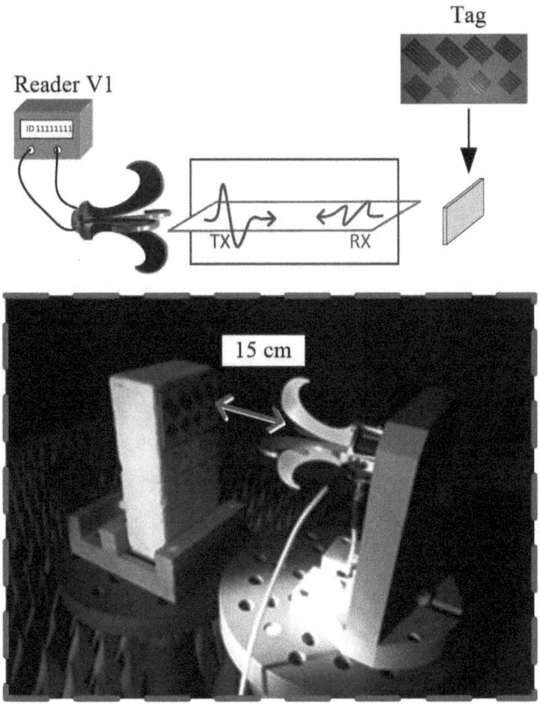


**Figure 3.12.** *Measurement bench specified to the antenna-tag configuration using the cross-polarization principle. The two tags are placed in turn at a distance of 15 cm, and first measured with a test bench based on a high performance DSO and then with the proposed reader*

The measurement results for both test benches, tags 1 and 2, in the time domain are shown in Figures 3.13(a) and (b), respectively. The measurements were executed with an averaging factor of 128 for the reader and of 64 for the DSO. A calibration process was previously executed to take into account the real delay of the reader's delay generator. In fact, it is composed of two chips connected in series, and the speculation of a tunable delay at an exact step of $\Delta T = 10$ ps is pretentious. The calibration process was based on a comparison between known measured signals with the DSO and the reader. The calibration results show a $\Delta T$ variation between 9 and 11 ps depending on the setup delay of the delay generator.

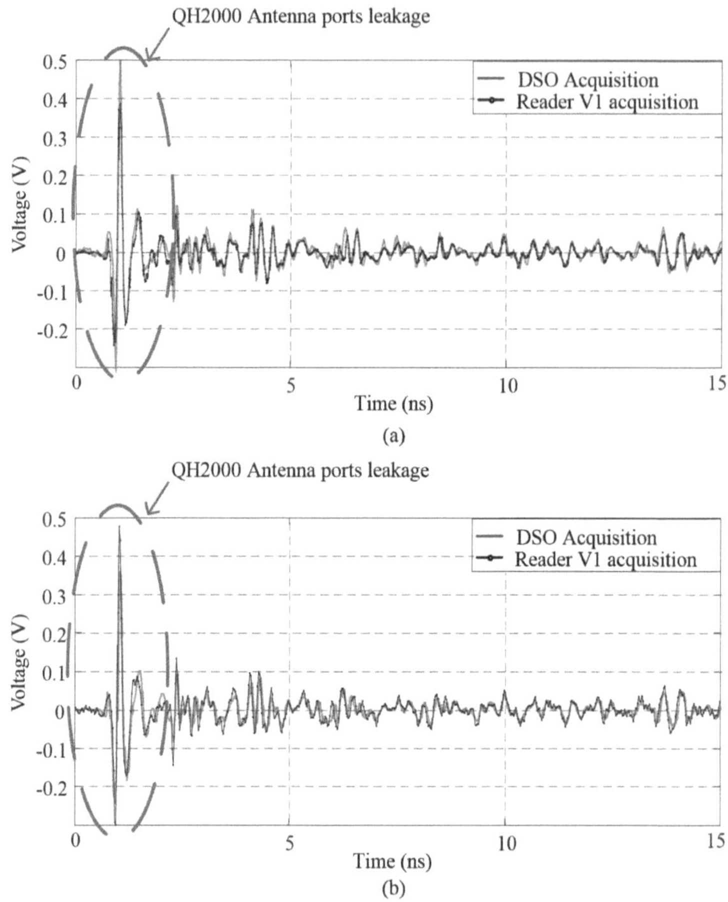


**Figure 3.13.** *Measurement results of chipless tag responses in the time domain (the corresponding measurement bench is shown in Figure 3.12). The DSO measurements are executed with an average of 64. Measurement of the proposed reader is performed with an average of 128: a) tag 1 and b) tag 2. For a color version of this figure, see www.iste.co.uk/garbati/chipless.zip*

The measurements shown in Figure 3.13 represent the tag measurement from which the background measurement is not yet subtracted. The higher voltage contribution at about 1 ns represents the leakage between the antenna ports. It represents the permanent contribution to the reader measurement.

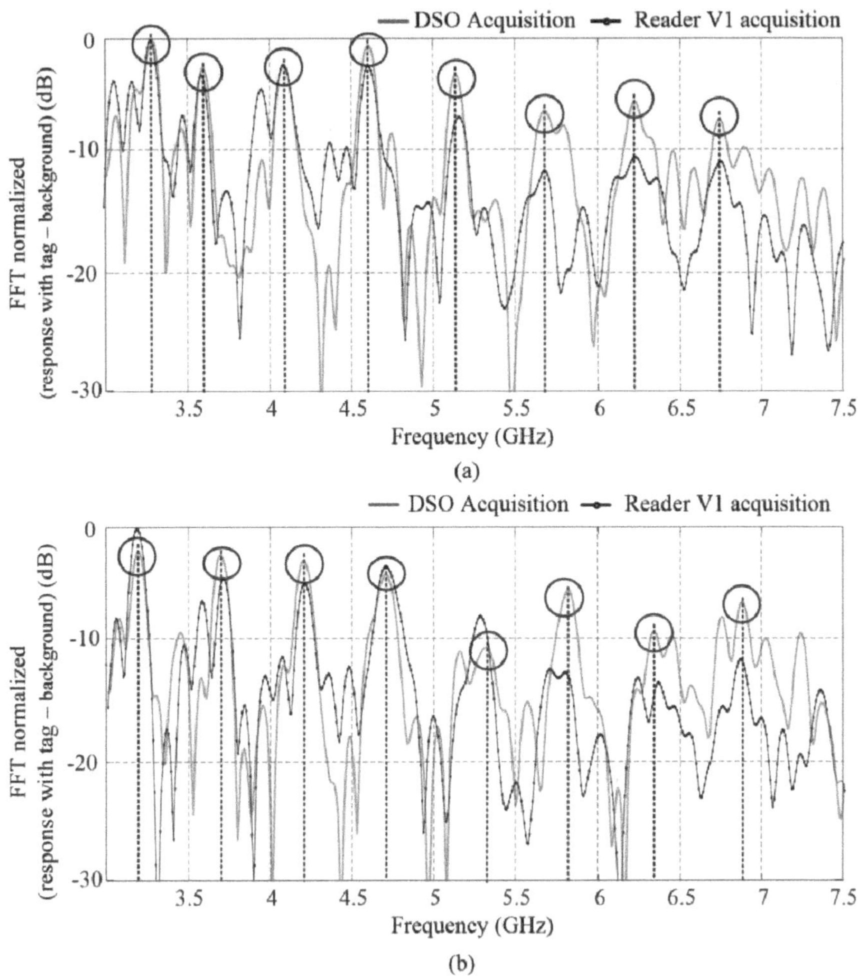


**Figure 3.14.** *Measurement results of tag responses in the frequency domain: a) tag 1 and b) tag 2. For a color version of this figure, see www.iste.co.uk/garbati/chipless.zip*

In Figure 3.14(a), the measurement results of tag 1 in the frequency domain is shown, and in (b) that of tag 2 after processing. For tag 1, the proposed reader shows the correct measurement for all of the resonant frequencies, where the DSO peak positions coincide with the measurement results of the reader. However, for tag 2, the reader did not correctly decode the information carried on by the sixth and seventh peaks. It is worth noting that even if the DSO correctly located the position of these two peaks, a decoding uncertainty is introduced due to the presence of a closer resonance with considerable power.

The obstruction of reading higher frequency resonances from the two test benches could have been overcome if the tag distance had been reduced by a few centimeters. In addition, the pulse generator Hyperlabs HL9200 has a lower power at higher frequencies, as shown in Figure 3.15(b), with the PRF of 1 MHz. Its response in the time domain with respect to a 50 Ω load is shown in Figure 3.15(a).

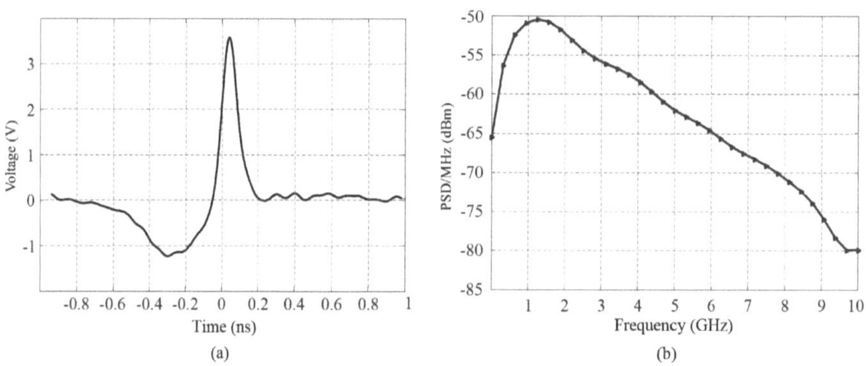


**Figure 3.15.** *Measurement of the pulse generator Hyperlabs HL9200 with the DSO in (a) the time domain and (b) the frequency domain with a PRF of 1 MHz*

## 3.7. Conclusion

In this chapter, a chipless RFID reader based on the IR-UWB approach was presented. It exploits the flexibility of an FPGA with its evaluation board to reduce the complexity involved in designing the reader hardware. It has a reading time in the order of several seconds, strongly depending on the averaging factor. The reader was validated in an anechoic chamber with measurements of two frequency-coded tags at a distance of 15 cm.

The acquisition suffers from synchronization problems inside the FPGA, which will be addressed in the following chapter by introducing a second reader version. Several aspects such as the reading flexibility (acquisition frame duration, frame position, equivalent sampling period $\Delta T$ and averaging), the reading time and the sampling clock period jitter will be optimized. The latter aspect is one the main characteristics that must be optimized in order to improve the reading performance.

# 4

# Optimized IR-UWB Chipless RFID Reader

## 4.1. Introduction

This chapter presents an optimized IR-UWB chipless RFID reader. Compared with the reader version proposed in the previous chapter, its major progress is characterized by the reduction of the reading time, with a factor of about 1,000, and the reduction of sampling noise. It also offers flexibility in terms of reading settings, such as acquisition frame position, length and equivalent sampling rate.

In the first part, a theoretical introduction to sampling noise and quantization noise is provided. In the second part, the reader hardware architecture is described in more detail, focusing on the characteristics that improved the reader performance. To facilitate the understanding of the chapter, some particular hardware and software aspects are provided in Appendices 3 and 4. Finally, measurements of chipless tags in practical environments are shown. This chapter is organized as follows:

– section 4.2 presents a theoretical introduction to ADC sampling and quantization noise;

– section 4.3 presents the hardware architecture of the reader, and demonstrates the reduction of its sampling noise compared with the reader version proposed in Chapter 3;

– section 4.4 describes the reader FPGA architecture, focusing on the techniques used to reduce the reading time of the reader and to resolve the synchronization problems addressed in Chapter 3;

– section 4.5 gives the specifications of the reader;

– section 4.6 presents the realized reader power supply board;

– section 4.7 shows some frequency-coded tag measurements;

– section 4.8 compares the reading accuracy of a frequency-based reader with an IR-UWB-based reader, which is assumed to be affected by different jitter levels;

– section 4.9 draws the conclusion of this chapter.

## 4.2. ADC noise theory

The reader proposed in Chapter 3 has an AD conversion block that consists of a UWB S/H amplifier followed by a driver and a low-cost ADC. The AD conversion block may be considered a unique ADC presenting a UWB signal at its input. The ADC block of a UWB chipless RFID reader may either have a low frequency signal as input for the frequency-based approach (see Chapter 2) or have a UWB signal for the IR-UWB approach.

All the analog-to-digital conversion processes degrade the SNR ratio at its input. There are different sources of noise that may play a major or minor role, depending on the ADC characteristics as well as on the input signal in terms of voltage range and frequency contents. An ADC can be considered an S/H amplifier in series with a quantizer, as shown in Figure 4.1. The S/H amplifier is responsible for the sampling of the analog input in the time domain, which ideally consists of a switch controlled by a clock and a capacitor to follow the input signal voltage. The switch commutes according to clock events such as rising and falling edges. It closes when falling events occur and opens with rising events. When the switch is closed, the capacitor follows the analog input signal. At the rising clock event (sampling), the switch opens and the capacitor is in a memory state. The tension of the capacitor is constant (holding) until the switch is closed again. During the memory state (holding), the capacitor value is converted into digital symbols by the quantizer, where the input/output characteristic can be represented by a staircase function (see Figure 4.1).

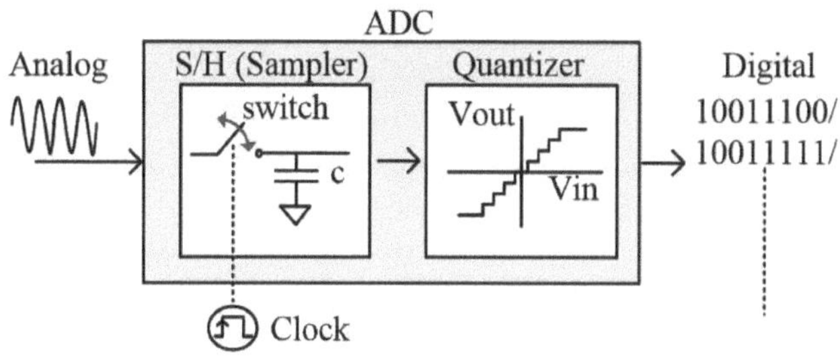


**Figure 4.1.** *Block schematic of the ADC. It is composed of an S/H amplifier in series with a quantizer. The former is responsible for sampling noise and the latter for quantization noise*

The output of the ADC has an SNR lower than its input. This is mainly due to the error introduced by the sampling and quantization processes. The sampling process adds sampling noise, which is correlated with the jitter of the sampling clock and the aperture time uncertainty of the ADC. The quantization process presents a quantizer with a nonlinear input/output characteristic and therefore introduces higher order noise, called quantization noise. For a frequency-based chipless RFID reader, the useful ADC input signal is almost DC and the SNR is mainly influenced by quantization effects. An ideal ADC with a staircase quantizer has a maximum SNR output limited to:

$$SNR\ (dB) = 6.02N_{ADC} + 1.76\ dB, \qquad [4.1]$$

where $N_{ADC}$ is the resolution (number of bits) of the ADC. Equation [4.1] assumes a sinusoidal input signal with an excursion equal to the full-scale ADC input, the ADC clock and the sinusoidal input period with an irrational ratio, and the clock frequency that is approximately equal to the Nyquist rate [BEN 48]. The SNR of an ADC can be improved by increasing $N_{ADC}$ (equation [4.1]) and by exploiting techniques such as oversampling and digital filter implementations. In the case of the UWB chipless RFID system based on IR-UWB, the ADC input signal is wideband and composed of the coupling between the transmitting and the receiving antennas, and the structural and antenna (tag) mode signals. The tag ID is in the tag mode frame for a frequency-coded tag, which is the least powerful part of the ADC input signal. Thus, to reduce the quantization noise in the antenna (tag) mode, it is important to select an ADC with a large number of bits and to amplify the antenna (tag) mode ideally to match it with the ADC input full-scale range.

The sampling noise depends on the ADC's hardware characteristics, sampling clock jitter and input frequency contents. If the input of the ADC is a low frequency signal, then the sampling noise may be negligible, which is the case for the frequency-based chipless reader. For an IR-UWB chipless reader, sampling noise is important because the ADC input is no longer a low frequency signal. For simplicity, we consider a sinusoidal ADC input, which is the result of the following stationary process:

$$V_{IN}(t) = \sin(\omega t + \phi) \qquad [4.2]$$

where $\phi$ is a random variable uniformly distributed between [-$\pi$, $\pi$]. This process is parametric, stationary and ergodic. Once this process is sampled by the S/H amplifier of the ADC (see Figure 4.1), it becomes:

$$V(k) = \sin(\omega(Tk + \gamma) + \phi) \qquad [4.3]$$

where T is the sampling clock period, and from the central limit theorem, $\gamma$ is a Gaussian random variable with zero mean and k is the integer. It considers the jitter noise of the ADC clock combined with the ADC aperture uncertainty, which generally has a minimum influence. The error $N(k)$ introduced by the $\gamma$ can be approximated with:

$$N(k) = \cos(\omega(Tk) + \phi)\omega\gamma \qquad [4.4]$$

This process is stationary; indeed, the two random variables $\phi$ and $\gamma$ are independent, and the fact that their mean values and autocorrelation are invariant with respect to timing translations can be easily demonstrated. The ratio between the power of the processes in equations [4.4] and [4.3] is equal to:

$$SNR = -10\log_{10}(\omega^2\sigma_t^2) \qquad [4.5]$$

where $\sigma_t^2$ is the variance of $\gamma$, which can be approximated with the variance of the sampling clock jitter. From equation [4.5], the SNR decays to 20 dB/dec compared with the input signal frequency. In the case of the UWB input signal (IR-UWB chipless reader), the calculation of the SNR can be found in [SMI 17], which is equal to:

$$SNR = -10\log_{10}(4\pi^2\sigma_t^2(f_0^2 + \frac{BW^2}{12})) \qquad [4.6]$$

where $f_0$ is the central frequency of the UWB input signal and BW is its bandwidth of –3 dB. For $f_0$ > 10 BW, the bandwidth of the signal can be neglected.

Only the jitter of the ADC sampling clock has been considered in this analysis. In the case of an IR-UWB chipless reader, the pulse generator is triggered by a signal, such as a low frequency clock, which is also affected by a jitter. If its jitter is uncorrelated with that of the ADC sampling clock, then its variance can be summed at $\sigma_t^2$. This is the case for the reader presented in the last chapter and the current optimized version presented here.

In the last reader version presented in Chapter 3, the estimated sampling clock period jitter standard deviation was of about 7.34 ps, while the reader input bandwidth was between 1 and 11 GHz. Substituting these values into equation [4.6], we obtain an SNR of 10 dB. With respect to the first reader version presented in Chapter 3 and realized with the project *Gravit*, the sampling clock jitter was even larger (about 13.22 ps). This standard deviation gives an SNR of about 5 dB; note that at this jitter has to be summed the jitter of the pulse generator trigger. This will result in an even worse SNR result.

The averaging process helps to increase the SNR, but at the expense of a higher reading time. Much care was spent in the design of the optimized reader presented here to ensure an ultra-low jitter. It may be considered as a cheap manner of improving the reader performance. Indeed, the jitter that affects a signal is directly correlated with the signal integrity. It is caused by different sources of noise: power supply variations, thermal noise, interference between nearby circuits and loading mismatching [MIT 09]. Therefore, the hardware architecture, the layout of the RF front-end acquisition board and the selection of the components can be optimized without an increase in the reader components' cost. On the one hand, for the frequency-based chipless reader, the error induced by the ADC can be neglected, and on the other hand, the VCO phase noise reduces the overall reader performance, decreasing its reading frequency resolution.

The reader presented here is introduced in the next section, and the RF front-end board schematic is described along with the solutions employed to reduce the jitter of the S/H amplifier sampling clock and the jitter of the pulse generator input trigger.

## 4.3. Reduced reader jitter: implemented hardware solution

According to equation [4.5], the sampling noise reduction is fundamental in simplifying the reading of the tags' information at higher frequencies. In this case, a lower reader averaging factor may be used to read the tags in order to reduce the reading time.

The block schematic of the proposed optimized reader is shown in Figure 4.2 and its photograph in Figure 4.3. The schematic shows two clocks, one at 73.333 MHz and the other at 125 MHz. The 125 MHz clock is used as the sampling clock for the S/H amplifier and also as one input of the ultra-low jitter AND gate. To avoid triggering the pulse generator continuously, the other input of the AND is connected to the FPGA in a clock gating configuration. As shown in Figure 4.3, the redesigned RF front-end board holds both the 125 MHz clock generator and the AND gate. The FPGA clock is increased from 67 to 73.333 MHz to speed up the UART communication with the PC, as briefly explained in section 4.4.1.1.

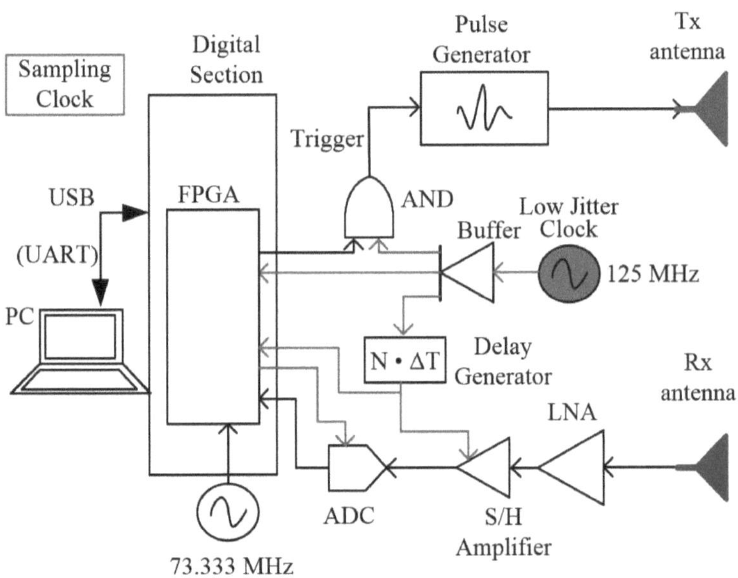


**Figure 4.2.** *An architecture of the optimized reader. It has a bi-static configuration, which is composed of three parts: a UWB pulse generator, a digital section and the receiving RF front-end. For a color version of this figure, see www.iste.co.uk/garbati/chipless.zip*

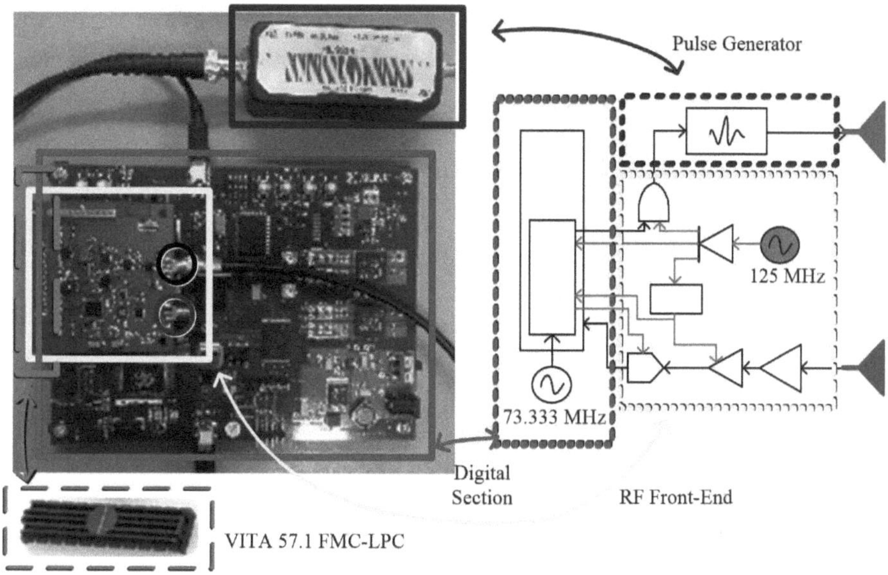


**Figure 4.3.** *A photograph of the optimized reader. The pulse generator is the HL9200 from Hyperlabs. The digital section is enclosed in a Xilinx SP601 evaluation board for the FPGA Spartan 6 XC6SLX16. The RF front-end board consists of four layers on an FR-4 substrate. For a color version of this figure, see www.iste.co.uk/garbati/chipless.zip*

A photograph of the RF front-end board is shown in Figure 4.4(a), where the main components are underlined. It is a four-layer board made on a low-cost FR-4 substrate. Its dimensions are 70 × 70 mm. The 125 MHz clock generator is the NBXDPA019 from ON Semiconductor. It is compatible with the PECL digital standard and has a period jitter standard deviation of 2 ps measured over 10 k cycles. The buffer is the CDCLVP2104 eight LVPECL outputs from Texas Instruments. The AND gate is the MC10EP05 from ON Semiconductor, which has an ultra-low jitter component compatible with both PECL and NECL (Negative ECL). It has a random clock jitter with a standard deviation of 0.2 ps measured between 1 and 3 GHz state rate variation. The delay generator is the MC100EP196, which is also used in the first reader version but in a more efficient package, the 32QFN (32 Quad Flat No-leads) instead of the 32LQFP (32 Low Profile Quad Flat Package). Because of the new sampling clock of 125 MHz, only one chip is required as a delay generator. A single delay generator chip has an internal delay dynamic variation of 10 ns, which is sufficient to cover the 8 ns period of the 125 MHz sampling clock.

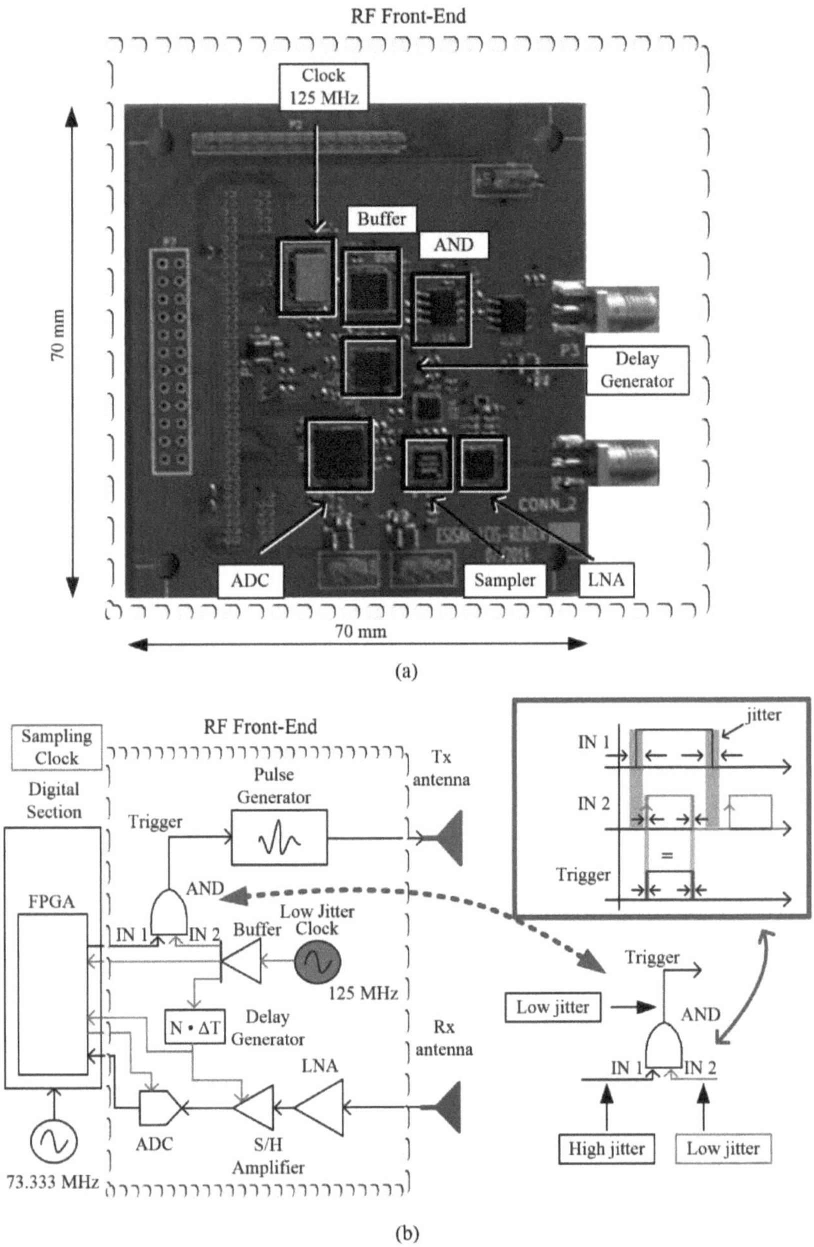


**Figure 4.4.** *a) A photograph of the realized analog RF front-end of the optimized reader. It is a four-layer board on an FR-4 substrate and measures 70 × 70 mm. b) Implemented schematic of the RF front-end board and AND gate operation. For a color version of this figure, see www.iste.co.uk/garbati/chipless.zip*

The pulse generator trigger input is not compatible with ECL standards; hence, a PECL–TTL converter chip was used after the output of the AND gate. In reception, the LNA (HMC753LP4E) and the S/H amplifier (1321TH) are the same as for the previous reader version, while the ADC is the ADS58B19 from Texas Instruments. It has 9 bits of resolution over 1.5 $V_{PP}$. It does not require a driver to be interfaced with the sampler outputs because its differential inputs front-end is composed of two analog buffers, and the common mode voltage is provided directly by the ADC. Thus, the two pins of the sampler output are connected to a resistor of 100 Ω for impedance matching, and are AC coupled with the ADC inputs.

The selected FPGA has a low cost and is not designed to provide signals with an extremely low jitter. Moreover, the connection between the SP601 evaluation board (digital section) and the designed RF board is made with a VITA 57.1 FMC-LPC connector (see Figure 4.3). It is not optimized for RF signal integrity, and therefore all those signals generated inside the FPGA, and travelling toward the RF front-end board, have an excessive jitter. This was the case for the previous reader version (Chapter 3), with the pulse generator trigger and the S/H amplifier sampling clock signals. In the proposed optimized reader, those two signals have a low jitter. This is because they are directly generated on the RF front-end board using the 125 MHz clock and the AND gate. The AND gate output is the pulse generator trigger, whereas its input is a high jitter FPGA output, denoted as *IN 1*, and the low jitter 125 MHz clock, denoted as *IN 2*. The two inputs provide a clock gating configuration. To keep a low jitter AND gate output, it is sufficient to assure the *IN 1* events (rising and falling edges) when the low jitter *IN 2* is stable *at* $0_2$, as shown in Figure 4.4(b).

The sampling clock period jitter of the S/H amplifier was measured as with the previous reader versions. It has a random period jitter over 1,000 cycles, with 2.45 ps of standard deviation (RMS) and a maximum peak to peak of 16.31 ps. It represents a huge improvement compared to the reader versions in Chapter 3 (7.34 ps RMS, p-p 67.37 ps). Applying equation [4.6] again, as in section 4.2, at the lower jitter standard deviation of 2.45 ps, an SNR of about 20 dB is obtained. This represents a huge improvement compared with the other reader versions. It was of 10 dB for the reader presented in Chapter 3 and 5 dB for the reader version made during the project Gravit. In the following section, the reduction of reader sampling noise is demonstrated.

### 4.3.1. *Temporal characterization of reader's RF front-end*

To stress the reduction of reader sampling noise, the reader pulse generator output was measured without any averaging, with the previous reader version (Chapter 3) and with the optimized version. The measurement results for the previous reader version are shown in Figure 4.5(a) and (b) for the optimized version. Between the pulse output and the RF reader input, some attenuators were interposed to avoid the reader front-end saturation. The same measurement performed with the high performance DSO Agilent 91204A is shown in red. The measurement in Figure 4.5(a) demonstrates a higher sampling noise than that in Figure 4.5(b), especially in the pulse rise and fall edges (higher slew rate). This is due to the jitter of the S/H sampling clock and that of the pulse generator trigger input.

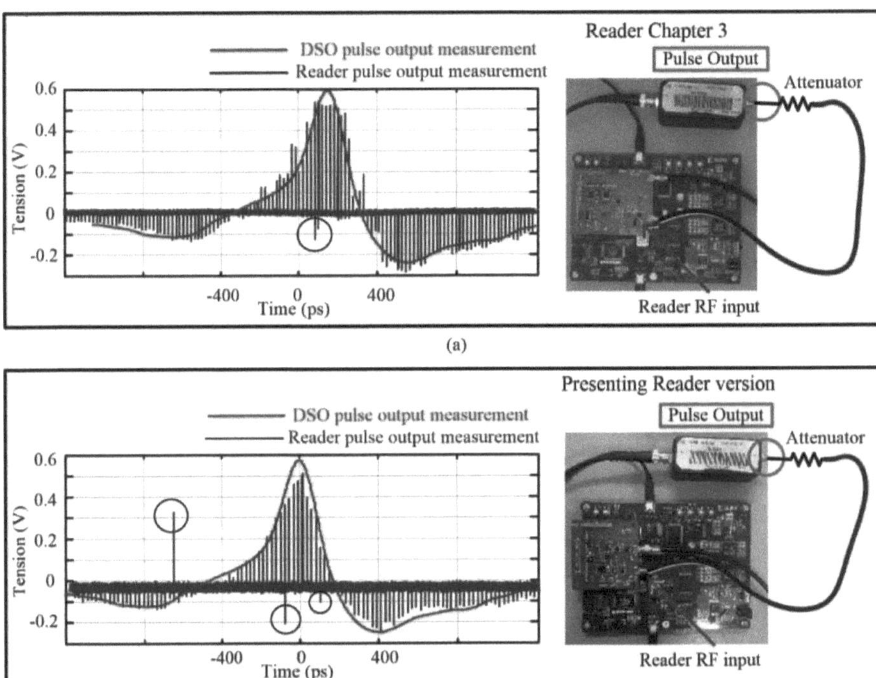


(b)

**Figure 4.5.** *Direct measurement of the pulse generator HL9200 from Hyperlabs, without averaging with the reader and the real time DSO Agilent 91204A. a) Measurement with the reader version presented in Chapter 3. b) Measurement with the reader presented here (low jitter). The higher sampling noise of (a) is evident, especially when the pulse has a higher slew rate. For a color version of this figure, see www.iste.co.uk/garbati/chipless.zip*

The sampling noise shown in Figure 4.5(b) is imperceptible, but both versions suffer from synchronization problems during the acquisition process. This is shown in the figure plots with black circles, where some acquired points indicate different timing positions. This type of problem has been observed in all the prototypes devised so far, which is solved by working on the implementation of the FPGA, as shown in the following section.

The new architecture of the FPGA will be discussed in the following section. The problem of synchronization and the reduction in the reading time are correlated with the internal implementation of the FPGA.

## 4.4. FPGA architecture

The block schematic of the FPGA embedded system is shown in Figure 4.6. Compared with the previously designed reader (Chapter 3), the new version makes use of an external flash memory. The embedded microprocessor system consists of the Microblaze 8.20a microprocessor connected to an internal BRAM of 8 kb with LMB bus in a Harvard architecture. The PLB bus is used to connect all the other peripherals: the 4-LED GPIO controller, the interrupt controller, the RS232 Uartlite controller to transfer information between the reader and the PC, the serial peripheral interface (SPI) for the external flash memory, the external DDR2 memory controller and finally the Reader_0 custom peripheral to control the reader acquisition. Considerable work has been done to optimize the FPGA architecture, especially for the custom block Reader_0, to eliminate synchronization problems during the acquisition process and to reduce the reading time.

The external flash memory allowed for an FPGA self-configuration at the reader start-up. With the previous reader version, the user has to download the bitstream into the FPGA at each start-up.

### 4.4.1. *Reducing reading time*

In this reader version, the reading time was optimized by working on the different aspects of FPGA implementation. Work has been carried out on the connection of UART to the PC in order to optimize the algorithm in coding the reader data to be transferred to the PC. In addition, low-level C functions were used, and the architecture of the Reader_0 custom peripheral was implemented with hardware averaging calculations.

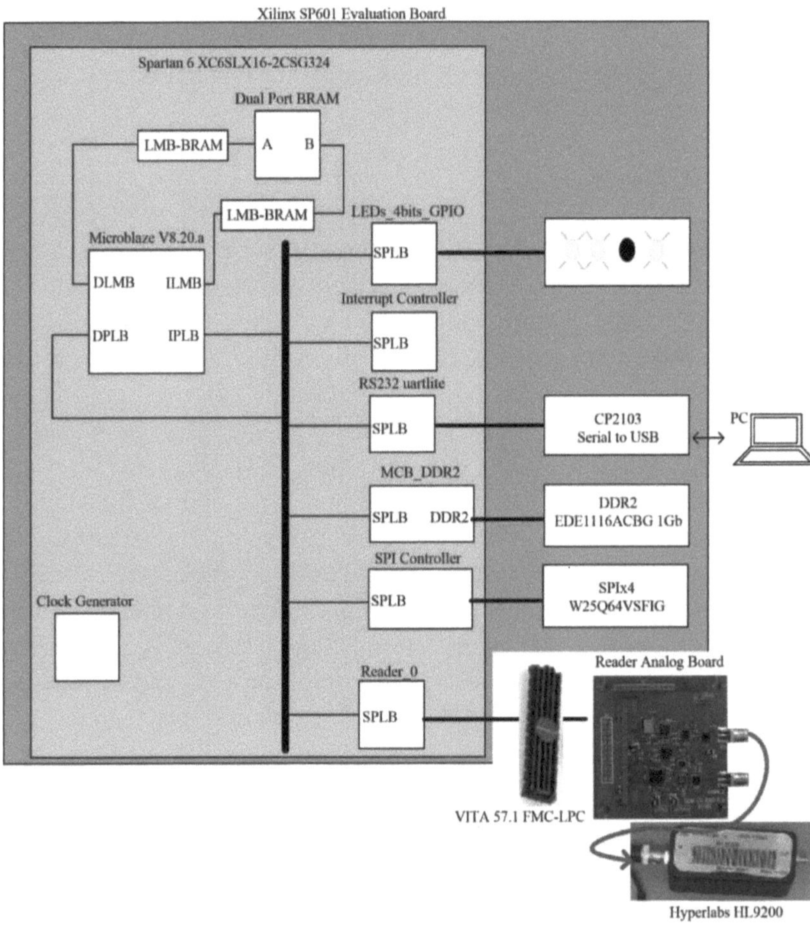


**Figure 4.6.** *A block schematic of the FPGA embedded system: with the main peripherals of the digital board SP601 used to implement the reader. The custom block Reader_0 can solely communicate with the analog board and trigger the pulse generator with the clock gating configuration. For a color version of this figure, see www.iste.co.uk/garbati/chipless.zip*

#### 4.4.1.1. *UART optimization*

The UART of the first reader version had a BR of 460.800 kb/s. In the new reader version, the BR was first increased from 460.800 to 921.600 kb/s. This modification sped up the communication, but put more constraint on the internal clock of the FPGA.

From the RS232 Uartlite data sheet [XIL 17a], the UART sampling clock is obtained using an integer counter at the FPGA main clock (master clock). Thus, in comparision with the aim of 14.746 MHz (921.600 kb/s × 16) for the serial communication, an error is introduced. A UART device may support a sampling clock error in the range of ±5% [XIL 17a]. In the case of the first reader version, the master clock was exactly 66.67 MHz instead of 67 MHz. The objective UART sampling clock was 7.3728 MHz (460.800 kb/s × 16), but the real UART sampling clock was 7.408 MHz (66.67 MHz/9) with an error of only 0.48% ((7.408–7.3728)/ 7.3728), which is within the ±5% limit. If the same master clock is applied to the larger BR of 921.600 kb/s, the UART sampling clock will be 13.334 MHz (66.67 MHz/5), with an error of 6.1% ((13.334–14.746)/14.746), which exceeds the limit of ±5%. Working with the clock wizard of the Xilinx Platform Studio software, a new master clock of 73.333 MHz was developed. This helps to reduce the error from 6.1 to only 0.54% ((14.666–14.746)/14.746), without violating the clock tolerance of the microprocessor (Microblaze 8.20a) and the attached peripherals. The use of a BR of 921.600 kb/s, instead of 460.800 kb/s, allows the time taken to transfer data to the main PC to be reduced by a factor of 2 when compared with the first reader version.

#### 4.4.1.2. *Data transfer to PC: algorithm optimization*

In the first reader version (Chapter 3), the ADC output resolution was 10 bits, and a complete acquisition frame, even with averaging, consisted of samples of 10 bits. Inside the application software written in C, the acquired frame was represented by the vector *pulse[i]*. For each word of 10 bits of *pulse[i]*, a maximum of 6 bytes was transmitted to the PC using the function *xil_printf* from the C Xilinx library. The American Standard Code for Information Interchange (ASCII) was used as a character encoding standard. The chunk of code of the application software in charge of it is represented as follows:

```
for(i = 0; i < num_rcvd; i++)

    xil_printf("%d\r\n", pulse[i]);
```

where *num_rcvd* is the length of the vector pulse, being equal to the number of acquired points inside the reader acquisition frame. At 10 bits (ADC resolution), it corresponds to an integer number between 0 and $2^{10} - 1$ (0–1023). Each can be represented in ASCII with a number of bytes ranging between 1 (0–9) and 4 (1,000–1,023), where each byte represents a digit. Each element of the vector *pulse* is the result of an averaging process. However, because they are integer 5, they do not have a fractional part and therefore are still represented by a maximum of 4 bytes in ASCII.

The argument of the function *xil_printf* also has the two ASCII characters *\r \n,* representing, respectively, carriage return and newline. These two characters are used to simplify the code development in C for the FPGA and in Matlab® for the PC. In fact, each element of the *pulse* vector is represented by a variable number of bytes (between 1 and 4); hence, a sort of ASCII character, to inform the Matlab® code about the bytes number for each *pulse* element, is required. In the new reader version, each element of *pulse* vector is represented by the same number of bytes, and thus the two ASCII characters *\r \n* are no more indispensable. The ADC has 9 bits instead of 10 bits; therefore, the ADC output is between 0 and 511 ($2^9 - 1$).

In the new reader version, each *pulse* vector element (0–511) is sent using only two ASCII characters. The *pulse* element is represented by 9 bits; therefore, the first ASCII character is in charge of the 6 bits from positions 5 to 0, while the second character is in charge of the last three bits from 8 to 6. Finally, the two characters (bytes) are manipulated to avoid the transmission of one of the 33 special ASCII characters, which are represented by the bytes with a decimal value ranging from 0 to 32.

To transmit the bytes to the PC, the function *XUartLite_Send* is used, which is provided with the RS232 UART component driver. It is a function at a lower level compared with *xil_printf* and so potentially faster. The *XUartLite_Send* function has three arguments, the pointer to the RS232 UART component instance, a pointer to the buffer of data to be transmitted and finally the number of bytes being sent. This function transmits the specified buffer of data using the UART in either the polled or interrupt driven mode. Because it is not a blocking function, it will return before the data has been sent by the UART [XIL 17b], and a sort of delay between each call of the function is needed. The function has been used with 8 as the third argument, thereby resulting in a faster vector transmission.

#### 4.4.1.3. *Hardware averaging implementation*

In the previous version, the averaging process was implemented by software. Considering an averaging factor of *N*, for each of the *N* acquisitions, the content of the Reader_0 internal FIFO memory was transferred to the external DDR2 memory, and finally the microprocessor (Microblaze) was used to perform the averaging. The whole process took different seconds, where only the transfer process of the data from the Reader_0 internal FIFO to the DDR2 counts for 140 ms at each acquisition.

In the newer reader version, a hardware averaging is directly implemented inside the Reader_0 block. The content of Reader_0 memory is transferred to the DDR2 only once, which is at the end of the acquisition, and no averaging process is required by the microprocessor (Microblaze 8.20a).

The Reader_0 block developed represents a big custom peripheral with its 387 slices over 2278 available and 8 BRAM (18 kb) blocks over 32 available. By comparison, the microprocessor Microblaze 8.20a, also implemented in the FPGA chip, counts for 882 slices, with the optional Barrel Shifter, integer multiplier, enabling optional Machine Status Register Instructions (MSRI) and pattern comparator. The same peripheral of the first reader version occupies 163 slices, which are less than half slices of the actual peripheral (387). To realize the Reader_0 block, the tool ISE Project Navigator within the Xilinx ISE design Suite 13.2 is used. It allows for simulation at all implementation levels, from behavioral to post-place and route, and provides a schematic editor useful for a complex design. The Reader_0 custom block of the FPGA hardware implementation (see Figure 4.8) consists of a User Logic unit and a PLB slave interface as with the previous reader version. The User Logic is realized with a hierarchical strategy, which consists of 10 main blocks.

To speed up the reading process, the User Logic is composed of three FIFO memories, but only one in the first reader version. These memories are FIFO_FAST_SAMPLE_PACKAGE, FIFO_FRAME_PACKAGE and FIFO_STORAGE_PACKAGE. The first is a small memory of 32 words of 9 bits, which is directly connected to the ADC output. The second is capable of 4,000 words, while the latter shows the same capacity of 4,000 but with words of 20 bits. The use of three memories allows for a fast hardware averaging implementation and assures a simpler Xilinx bitstream generation.

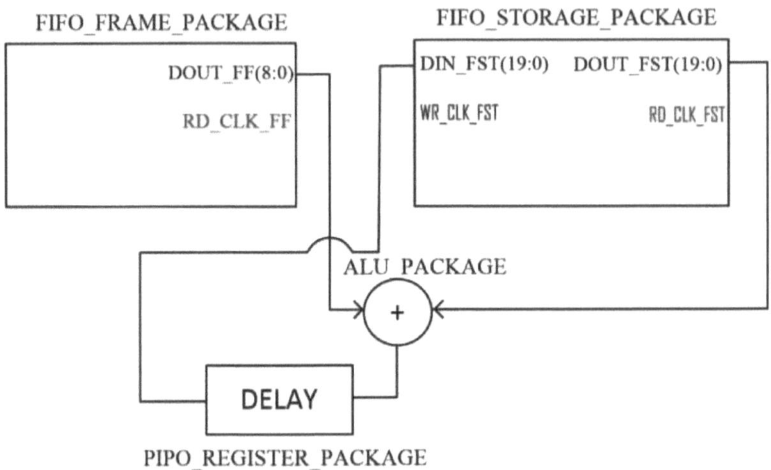


**Figure 4.7.** *Description of the fast averaging principle. The content of the FIFO_FRAME_PACKAGE is added to that of the FIFO_STORAGE_PACKAGE N times, where N is the averaging factor. After the process is terminated, the correct 9 bits over the 20 bits output are selected depending on N, in a shifting position-based division*

The part of the User Logic that takes care of the hardware averaging is shown in Figure 4.7. The third memory The FIFO_STORAGE_PACKAGE is used in collaboration with the ALU_PACKAGE and PIPO_REGISTER_PACKAGE. For an averaging of N, with N>1, a tag response measurement is first executed and memorized in the FIFO_FRAME_PACKAGE after its content is transferred to the FIFO_STORAGE_PACKAGE as it is. Then, for the N-1 remaining acquisitions, the content of FIFO_FRAME_PACKAGE is added to the content of FIFO_STORAGE_PACKAGE and together memorized in FIFO_STORAGE_PACKAGE. Therefore, the FIFO_STORAGE_PACKAGE is read and written at the same time.

The PIPO_REGISTER_PACKAGE is a Parallel Input Parallel Output (PIPO) register [XIL 11a]. It is used as a delay block, to compensate for the initial latency between reading and writing processes of the FIFO_STORAGE_PACKAGE, when executed simultaneously [XIL 11b]. The FIFO_STORAGE_PACKAGE memory accounts for words of 20 bits, while of 9 bits for the other two memories. This is because during a measurement process, with a predefined averaging of N, each word of the FIFO_STORAGE_PACKAGE contains the summation of N words of 9 bits. Thus, the maximum averaging factor depends on the length of

FIFO_STORAGE_PACKAGE words. In binary representation, the multiplication of a number by 2 and the division by 2 corresponds, respectively, to its shifting toward the left and the right of one position. The maximum possible averaging factor is therefore equal to 2048 (($2^{20-9}$)), limited in practice to 512 by software and restricted to a power of 2. For instance, if the averaging factor is 16, which is equal to $2^4$, from each FIFO_STORAGE_PACKAGE word, the 9 bits representing the averaged measurement to transfer to the PC are those from positions 4 to 13 (4 – 9 + 4).

#### 4.4.1.4. *Conclusion*

The reading time of the version of the reader presented here accounts for a reducing factor of about 1,000 compared to the previous version. As a result of the flexibility of the proposing reader, the reading time depends on the chosen parameters. The minimum reader time was about 6 ms, without averaging, and about 50 ms, with an averaging of 64.

### 4.4.2. *Solving acquisition synchronization problems*

The connection between the main components of the reader RF front-end board and the custom block Reader_0 inside the FPGA chip (Spartan 6) is shown in Figure 4.8. Compared with the first reader version, it shows a more complex schematic, where two clocks are involved, the master clock inside the FPGA at 73.333 MHz and the clock of 125 MHz generated from the analog board (RF front-end).

The clock of the ADC need not be of low jitter because, besides the trigger of the pulse generator, the reader sampling noise depends solely on the clock of the S/H amplifier, as demonstrated in section 4.3. The output of the S/H amplifier is a constant value for a complete clock period, so the fluctuations of the ADC sampling event caused by the jitter do not create an error in the acquired signal. From Figure 4.8, the Reader_0 block is composed of a PLB slave interface and a User Logic, where the first allows the communication between the User Logic and all the other peripherals connected to the same PLB bus. The output of the delay generator, which is the sampling clock, is connected to the S/H amplifier and also to the FPGA through the User Block. The connection with the FPGA was believed to simplify the design. However, finally, the User Block is optimized to avoid the use of this signal, as underlined in Figure 4.8. This optimization may reduce the sampling noise in a future reader version because the buffer at the output of the delay generator would no longer be needed. It will also reduce the reader's realization cost, for the absence of the buffer as well as of the digital standard converter PECL–TTL chip. The use of the digital standard converter is essential to increase the dynamic excursion of the signals crossing the VITA 57.1 FMC-LPC connector to reach the FPGA. The connector is not considered for high frequency signals, and at 125 MHz, they are attenuated considerably.

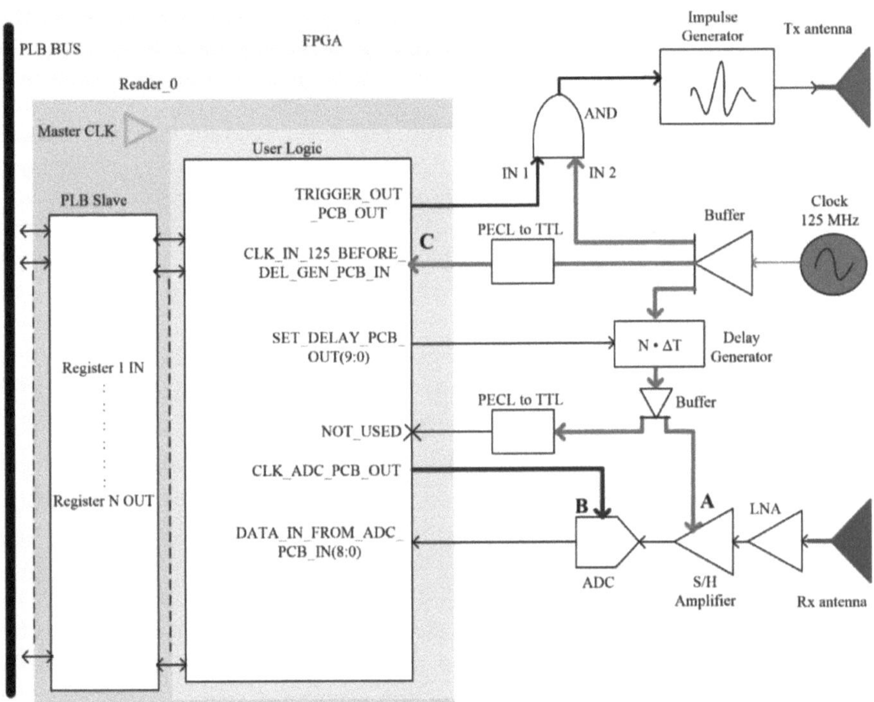


**Figure 4.8.** *Connection between the main components of the reader RF front-end board and the custom block Reader_0 inside the Spartan 6 FPGA chip. Compared with the first reader version, it deals with many synchronization problems to improve the tag reading quality. For a color version of this figure, see www.iste.co.uk/garbati/chipless.zip*

Contrary to the previous reader version (Chapter 3), many constraints about synchronization of different signals have been added to the new one to improve reader performance. According to Figure 4.4(b), a precise temporal relation exists between the two AND gate inputs, *IN 1* and *IN 2*, as discussed in section 4.3. From Figure 4.8, the ADC has the clock sourced by the FPGA and can ideally sample its analog inputs only when the S/H amplifier output is stable. The ADC output (DATA_IN_FROM_ADC_PCB_IN(8:0)) has to be memorized inside the User Logic memory with a proper clock. As shown in Figure 4.5, all these synchronization problems degrade the reader acquisition if not accurately handled.

The cooperation between the digital board (Xilinx SP601) and the RF front-end is significant, and the design of the User Logic becomes quite complex. In addition, the delay of different signals connecting the FPGA to the RF front-end is not characterized, which is caused by using a commercial digital board (SP601). However, the FPGA provides different resources to help resolve synchronization problems such as those considered in this project.

The ADC sampling clock is sourced by the FPGA, which is obtained from the sampling clock of the S/H amplifier before being delayed by the delay generator (see Figure 4.8). During an acquisition, the dynamic internal delay of the delay generator varies with minimum time steps of $\Delta T = 10$ ps to cover the sampling clock period which is 8 ns (1/125 MHz). Thus, the sampling clock signals of both the ADC and the S/H amplifier have a variable phase relation during the acquisition. At the beginning of an acquisition process, if the two clocks are out of phase, as shown in Figure 4.9(a), their phase relation will be modified during the acquisition at each update of the internal delay of the delay generator. As shown in Figures 4.9(b) and (c), for certain delay generator values, the two clock rising edges of both the ADC and the S/H amplifier are close, which may result in an ADC with no optimal operation acquisition.

Considering the case shown in Figure 4.9(b), the S/H amplifier sampling clock rising edge occurs before that of the ADC. The output of the S/H amplifier may be considered a step signal. If the two rising edges are very close in time, the ADC limited analog input bandwidth cannot follow the sudden S/H amplifier output variation at the sampling clock rising edge event. In this case, an analog ADC data setup time can be defined. Equation [4.7] is the classical step response relationship between the filter output rise time (10/90 %) and the bandwidth [DEL 98]:

$$Rise\ time = \frac{0.35}{Bandwidth}. \qquad [4.7]$$

The ADC input analog bandwidth is 550 MHz. Therefore, from equation [4.7], the analog setup time should be at least 0.63 ns.

The ADC samples its analog input at the clock rising edge with an aperture delay of about 0.8 ns. This can be viewed as an ADC's analog data hold time with respect to the ADC's clock rising edge, which must not be violated. From the ADC specifications, the recommended analog hold time is 1.2 ns. The S/H amplifier is able to actualize its output at each rising edge of its sampling clock, with an aperture delay of only 55 ps. Therefore, the S/H amplifier's sampling clock rising edge must respect the ADC's analog hold time if it is subsequent to the ADC's sampling clock rising edge, as shown in Figure 4.9(c).

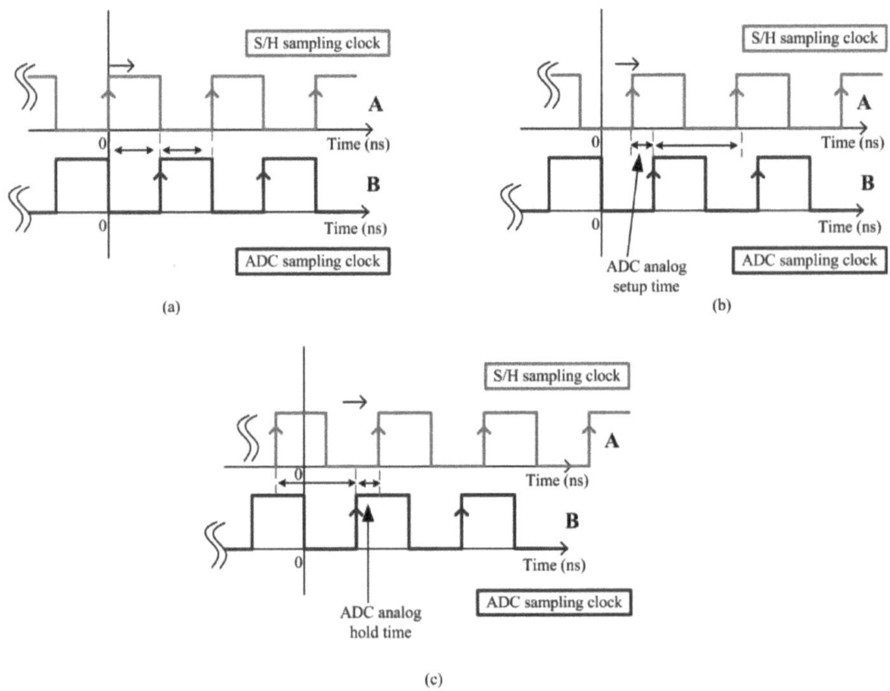


**Figure 4.9.** *The ADC sampling clock (blue lines) and the S/H amplifier sampling clock (orange lines) have a variable phase relation during the acquisition process. a) At the beginning of the reader acquisition, the two clocks are out of phase. b) The S/H amplifier sampling clock event still anticipates the ADC sampling clock event, but they are closer in time, and the ADC analog setup time may be violated. c) The S/H amplifier sampling clock event is subsequent with respect to the ADC sampling clock event. The two events are close, and the ADC analog hold time may be violated. For a color version of this figure, see www.iste.co.uk/garbati/chipless.zip*

In the FPGA User Logic implementation, the required dynamic delay of the delay generator (0–8 ns) is divided into three sub-frames: 0–2.55 ns, 2.55–6.4 ns, and 6.4–8 ns. From Figure 4.10(a), at the beginning of the first sub-frame, the ADC clock and the S/H amplifier clock are out of phase. After varying the dynamic delay to 2.55 ns, their phase relation changes according to Figure 4.10(a). To attain their first phase relation, the FPGA internal PLL utilities are exploited [XIL 15a]. In the first sub-frame, the minimum analog ADC setup time is 1.45 ns with a hold time of 4 ns. It does not violate the defined ADC setup time and hold time (0.63 ns and 1.2 ns, respectively).

For the second sub-frame of the dynamic delay of the delay generator (2.55–6.4 ns), the ADC sampling clock phase is varied by 180° using the same PLL, as shown in Figure 4.10(b). The minimum analog setup time is 1.6 ns with a hold time of 2.55 ns. For the last sub-frame 6.4–8 ns, shown in Figure 4.10(c), the ADC clock is again phase shifted by 180°, and the minimum ADC analog setup time and hold time are 4 and 2.4 ns, respectively.

The division of the dynamic delay generator frame in three sub-frames, with the variation of the ADC sampling clock phase at a step of 180°, assures a lower SNR and reduced synchronization problems. Other synchronization problems involve the User Logic block design. The internal memorization of the ADC output is optimized with an IODELAY2 block inside the FPGA, which provides a variable delay generator with a step of about 40 ps [XIL 15b]. It is used to properly delay the ADC output signal to respect the corresponding User Logic FIFO setup time and hold time.

The required temporal relation between the two AND gate inputs, *IN 1* and *IN 2*, was detailed in section 4.3 and shown in Figure 4.4(b). To properly gate the IN 2, for each trigger event the FPGA IN 1 signal is high ($1_2$) for one clock cycle, which corresponds to a duration of 8 ns. With respect to simple logic gate implementations, its output can be delayed by 4 ns, which corresponds to a half clock cycle (180°), to best fit the low jitter clock (125 MHz) gating requirement of Figure 4.4(b). In the following section, the same experiments described in section 4.3.1 are executed, so as to resolve reader synchronization problems during the acquisition process.

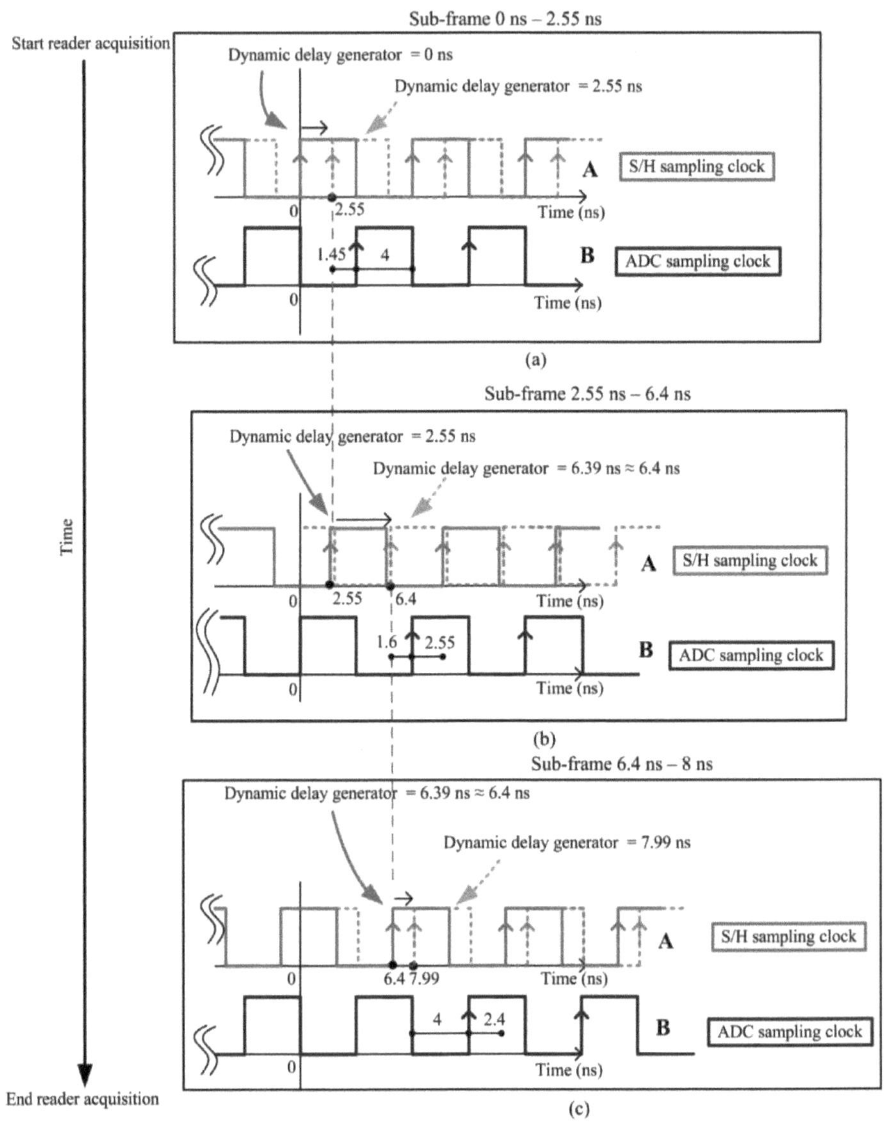


**Figure 4.10.** *Progress of the ADC sampling clock and the S/H amplifier sampling clock during a reader acquisition process. a) Progress of the delay generator dynamic delay between 0 and 2.55 ns. b) Delay generator dynamic delay between 2.55 and 6.4 ns. c) Delay generator dynamic delay between 6.4 and 8 ns. For a color version of this figure, see www.iste.co.uk/garbati/chipless.zip*

### 4.4.3. *Demonstration of the resolution of synchronization problems*

The same pulse output measurement shown in Figure 4.5 is executed, and the results are shown in Figure 4.11. This figure compares the reading performance with the version of the reader presented here, before and after resolving the synchronization problems addressed in this section in Figure 4.11(a) and (b), respectively. The measurement with the optimized User Logic in Figure 4.11(b) does not show any synchronization problem, contrarily to Figure 4.11(a) (points with black circles).

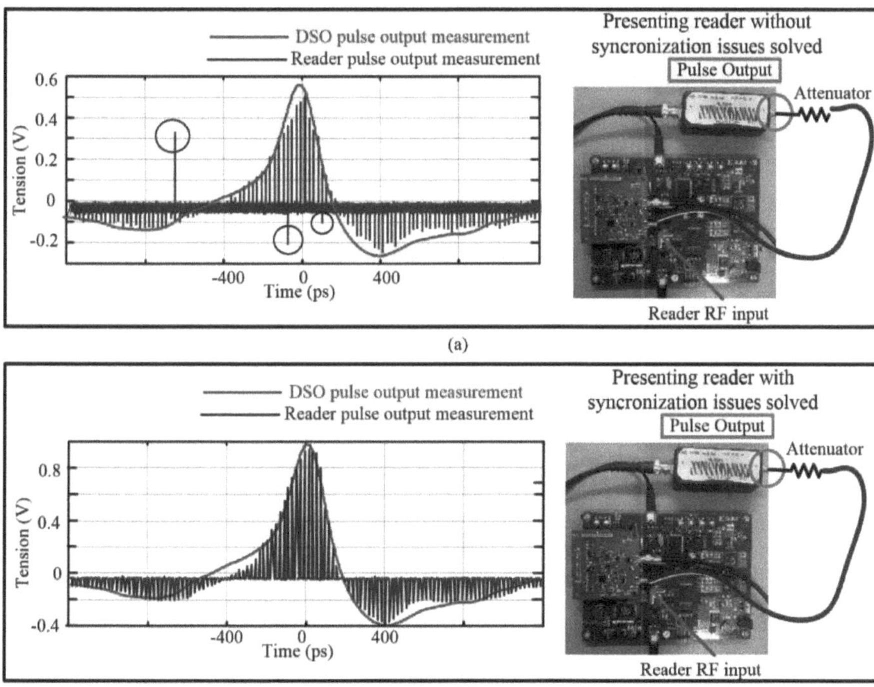


(b)

**Figure 4.11.** *Direct measurement of the pulse generator HL9200 from Hyperlabs without averaging with the reader and the real-time DSO Agilent 91204A. a) Measurement with the reader presented here before dealing with the synchronization problems (black circles). b) Measurement with the synchronization problems solved. For a color version of this figure, see www.iste.co.uk/garbati/chipless.zip*

## 4.5. Reader specifications

The reader costs and performance are given in Table 4.1. With respect to the previous version (Table 3.2, Chapter 3), it shows a limited reduction of the retail cost, from 2,350 € to 2,306 €. It should be noted that the cost of the Xilinx SP601 evaluation board has increased to 70 € since 2014. The S/H amplifier sampling clock period jitter has a variance reduced by three times. The reader shows flexibility in terms of acquisition frame position, length and equivalent sampling period (ΔT). This flexibility is obtained by working on the FPGA Reader_0 block architecture and using the FPGA application software.

| **Reader's retail cost after tax** | |
|---|---|
| **Component** | **Retail cost after tax (2013*/2017**)** |
| **RF front-end board components** | 886 € ** |
| **Digital board (Xilinx SP601 E.B.)** | 420 € ** |
| **Pulse generator (Hyperlabs HL9200)** | 1,000 € * |
| **Total cost** | 2,306 € |
| **Reading performance** | |
| **PRF** | 1 MHz |
| **Input bandwidth** | 1–11 GHz |
| **Acquisition frame duration** | 8–256 ns/step of 8 ns |
| **Frame position** | Variable |
| **Equivalent sampling period (ΔT)** | 10/20/30/40 ps |
| **Averaging** | [1–512] (power of 2) |
| **Reading time** | From 6 to 500 ms |
| **S/H amplifier sampling clock period jitter** | 2.45 ps RMS (16.31 ps p-p) over 1,000 cycles |
| **Reader sensitivity** | –60 dBm |

**Table 4.1.** *Components' retail cost after tax of the optimized designed reader presented here (without the antenna) and performance specifications*

The averaging factor is a variable between 1 and 512, with a power of 2. The reading time depends on the chosen acquisition parameters, with the minimum being about 6 ms and the maximum about 500 ms. Finally, the reader shows a sensitivity of –60 dBm; this value can be improved using an ADC with a better resolution at the expense of a higher reader cost.

The Matlab® application software for the PC is described in Appendix 3. It uses the Short Time Fourier Transform (STFT) algorithm, where in [RAM 16b], the STFT for REP-based UWB frequency-coded chipless RFID tags is studied.

## 4.6. Reader's power supply board

The reader's RF-front end board and pulse generator require diverse power supply points. The FPGA development board provides power pins on the VITA 57.1 FMC-LPC connector. The most useful reader requirements are the 12 V pins with a maximum current of 1 A, the 3.3 V pins with 3 A and finally the 2.5 V pins with 2 A. The required supply voltage for the RF front-end board is 3.3 V for the majority of the components, 1.8 V for the ADC, 5 V for the LNA and S/H amplifier, and −5.2 V for the S/H amplifier. Finally, 9 V is required by the pulse generator HL9200 from Hyperlabs. The 3.3 V is taken directly from the corresponding pins on the VITA connector, while the others are extracted from the 12 V pins with a developed board. A photograph of the alimentation card mounted on top of the RF front-end is shown in Figure 4.12. It is a two-layer board on a low-cost FR-4 substrate; the bottom layer is a ground plane without signal traces to reduce interference with the analog board underneath. The schematic of the board is shown in Appendix 4.

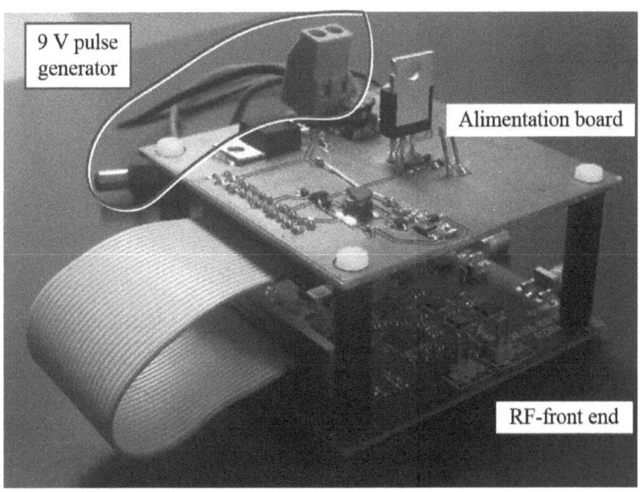


**Figure 4.12.** *Alimentation board of the reader mounted on top of the RF front-end board*

## 4.7. Reader's tag measurements

The reader was tested with intensive tag measurements. For comparison, the same UWB frequency-coded tags, used for testing the first reader version (see section 3.6 of Chapter 3), were used to test this optimized reader.

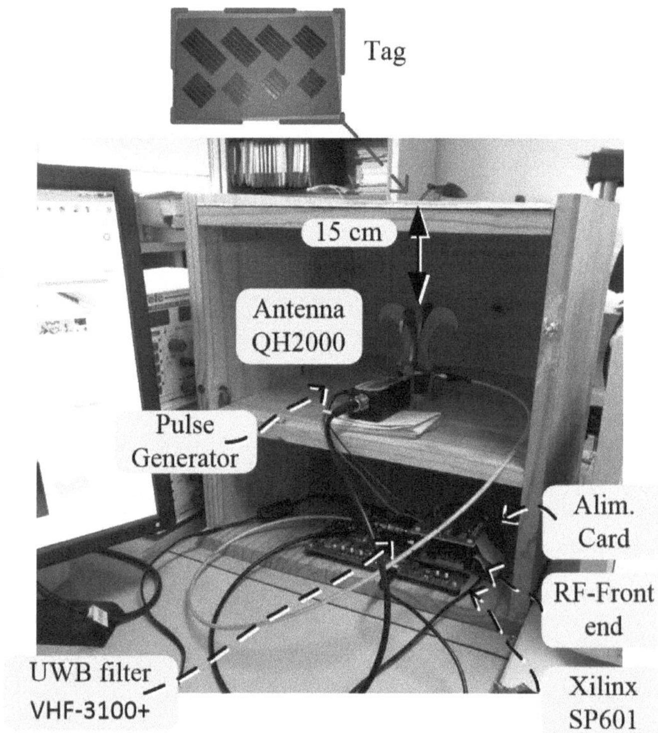


**Figure 4.13.** *Measurement bench for UWB frequency-coded tags in a practical environment for the reader and the DSO in a non-line-of-sight condition*

The measurements were performed with the reader and the DSO at a distance of about 15 cm from the tag. In Figure 4.13, a photograph of the measurement bench for the reader case is shown. It consists of a typical office scenario with metallic cabinets, chairs, tables and computers. The Satimo QH2000 dual-access dual-polarization antenna is connected to the reader to have a cross-polarization reading scheme. In reception, the UWB filter Mini-circuits VHF-3100+ was placed to eliminate the out-of-band noise such as the Wi-Fi at 2.4 GHz. In a real-time approach, even with averaging, the out-of-band signals may be filtered by software. In an equivalent time approach, this is not possible, and the out-of-band signals may add noise in the frequency band of interest. To clarify this point, it is useful to recall that in this approach, the sampling frequency is very low and of 125 MHz. It does not respect the Nyquist–Shannon theorem when used to acquire UWB signals that are in the band of 3.1–10.6 GHz. Therefore, for each acquisition, the out-of-band noise is replicated inside the band of interest. Although the averaging factor helps to reduce this noise, it is still important because of the low power response of the tags. For the DSO, the same setup was employed using the same pulse generator HL9200 from Hyperlabs and the antenna Satimo QH2000. From Figure 4.13, the tag is placed on top of a square of wood of a few millimeters of thickness with the antenna below. This setup creates a successful reading process in a non-line-of-sight condition, which is not possible with barcode technology.

The measurement results using the STFT algorithm described in [RAM 16b] are shown in Figure 4.14. The eight resonant frequencies match for both tags between the DSO and the reader. The tags can be decoded without any error.

The reader was shown in the Drupa 2016 event in Düsseldorf, Germany, from 16 June to 25 June, which is accredited as the biggest showroom in the world for printing technologies. The reader worked without any interruption for 10 hours a day, reading frequency-coded tags printed on a single-layer Powercoat PW230 paper substrate. A reader box was designed with the open access FreeCAD software and realized with a 3D printing machine. A photograph of the reader within its box during Drupa 2016 is shown in Figure 4.15. In the following section, the frequency-based and the IR-UWB approaches will be studied in terms of reading tag accuracy, focusing on the impact of the jitter for an IR-UWB system.

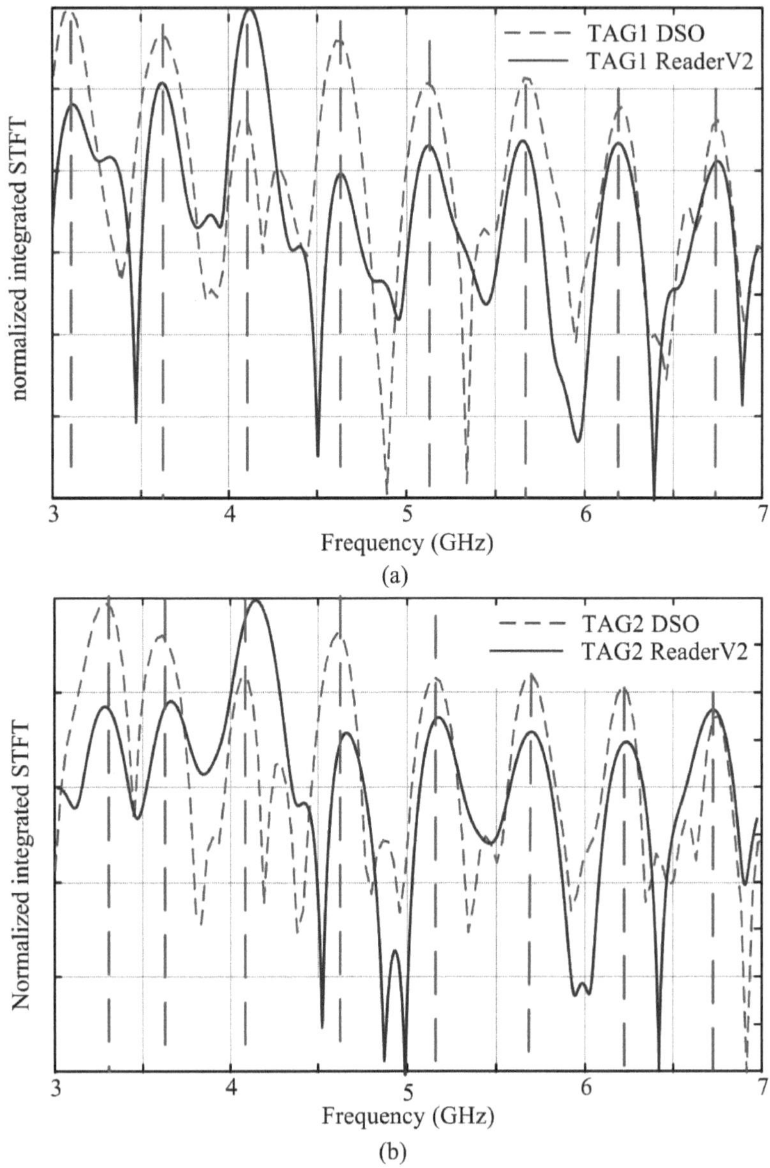


**Figure 4.14.** *Measurement results of two UWB frequency-coded tags, with the reader as a blue solid line and the DSO as a red dashed line. In the results for both (a) the first and (b) the second tag, the curves indicate the same resonant frequencies as expected. For a color version of this figure, see www.iste.co.uk/garbati/chipless.zip*

**Figure 4.15.** *A photograph of the reader within its designed box at the Drupa 2016 event in Düsseldorf, Germany*

## 4.8. Frequency-based reader versus IR-UWB with different jitter levels

In the first two chapters, the two frequency-based reader approaches, SFCW and FMCW, used in the literature for UWB chipless applications were introduced. A frequency-based approach provides an opportunity to perform more accurate measurement compared to an IR-UWB solution. This is because of the higher dynamic range (sensitivity) that can be obtained with such a solution. The interrogating signal sweeps around the frequency band of the tag, where the sweep time is defined at the reader level. Conversely, in IR-UWB systems, the interrogating signal is a short pulse in the UWB band, and the tag response significantly lasts for only a few nanoseconds (about 12 ns) [RAM 16b]. Owing to the short duration of the tag response, and its amplitude envelope, an IR-UWB system can perform tag reading with a limited frequency resolution compared to a frequency-based reader. Also, an IR-UWB system is affected by the sampling clock jitter and a higher input noise bandwidth. According to equation [4.5], the impact of the sampling clock jitter is more significant at a higher frequency, where, at small unwanted timing translations of the sampling clock event, it corresponds to a higher phase variation of the acquired signal. In contrast, an IR-UWB approach is easily compatible with UWB regulations, as shown in Chapter 2, and has a reduced reading time. For instance, the reader presented in this chapter can have a reading time as low as 6 ms, with an input bandwidth ranging between 1 and 10 GHz.

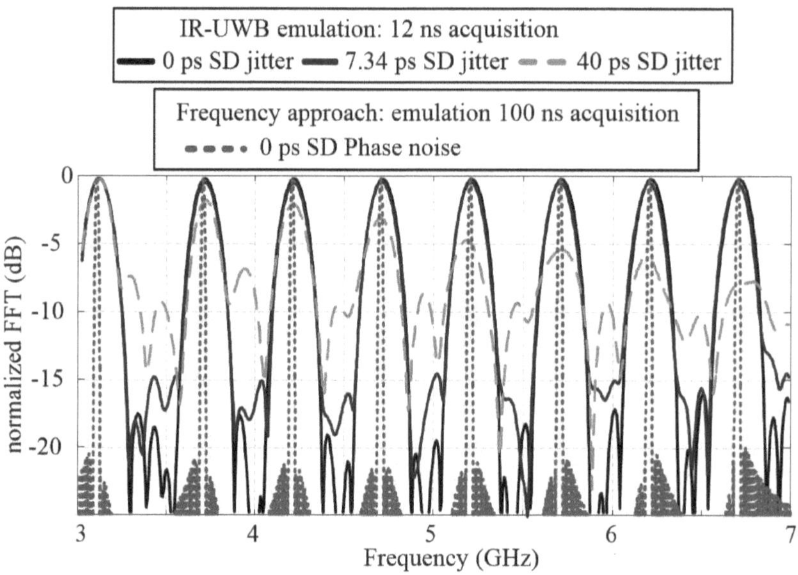


**Figure 4.16.** *Emulation of the frequency response of a chipless tag read with a frequency-based approach and with an IR-UWB reader working in equivalent time. The IR-UWB reader is affected by a jitter of 0, 7.34 and 40 ps. The tag with a longer response (100 ns) corresponds to an FMCW approach, where an IF bandwidth can be in the order of kHz. The shorter tag response (12 ns) corresponds to the case of an IR-UWB approach. For a color version of this figure, see www.iste.co.uk/garbati/chipless.zip*

An equivalent time reading system with different parameters has been emulated in Matlab®, and the results are shown in Figure 4.16. An ideal tag mode response, where the tag is assumed to have eight resonances between 3.2 and 6.7 GHz, has been emulated. The resonators are assumed to have an infinitive quality factor. Two different tag response durations (100 and 12 ns) are considered. In the case of 12 ns, the reading system is assumed to be affected by a jitter of 0, 7.34 and 40 ps, respectively. The 7.34 ps jitter is the jitter of the reader presented in Chapter 3, while the 40 ps jitter corresponds to the jitter of the IR-UWB radar for through-wall applications introduced in [LIU 13]. Comparing the effect of the two different signal durations without jitter, the larger one allows for a better reading frequency resolution. It may be translated with the aim of developing tags with a smaller inter-distance between resonances and therefore a higher bit capacity.

The larger duration corresponds to the case of a frequency-based interrogation, where the IF bandwidth can be in the order of kHz. The effect of the jitter variation for the smaller response duration is also studied. It represents a real tag signal response in the case of an IR-UWB-based interrogation system. The jitter augmentation affects the reading capability, as shown in the figure, where the distance between the main lobs of the tag's response (which are located around the tag's frequency resonances) and the secondary lobs is smaller. Moreover, the peak apexes corresponding to these main lobs decrease with frequency, and their frequency value and bandwidth also vary, especially at a high frequency.

A chipless tag acquisition with different reading systems, which are based on SFCW and IR-UWB with different performances in terms of jitter, is shown in Figure 4.17. The frequency domain chipless tag employed shows eight resonances between 3 and 7 GHz, which is the same as that used to characterize the readers so far. All the measurements were performed in an anechoic chamber to reduce the noise distinct from the jitter system as much as possible.

The measurement with the VNA N5222A from Keysight emulates a chipless reader based on SFCW. It is costly test equipment with performance superior to all the presented frequency-based chipless readers in the literature [KOS 12, POP 16b, PRE 10b]. The DSO 91204A measurement was performed with the Hyperlabs HL9200 pulse generator in an IR-UWB configuration. The *reader V1* indicates the reader version presented in Chapter 3, while the *reader V2* indicates the optimized reader presented here.

As expected, the measurement with the VNA presents the best performance with a higher frequency resolution. In addition, compared to the other measurements (IR-UWB configurations), all the peaks show a comparable power response. This is because, in case of IR-UWB, the pulse generator output power is not constant in all the analized UWB band. Nevertheless, the measurement performed with *reader V2* and the DSO is similar and completely successful. Finally, the first reader version *reader V1*, which has a larger jitter of 7.34 ps, presents the worst reading performance. It is not possible to correctly read the resonance positions at a high frequency, and the others present a shift in frequency of a few MHz. This drop in performance is mainly due to the effect of the system jitter, as well as to the noise generated by the inefficient application of the equivalent time algorithm because of synchronization problems. It can add noise to the entire tag's bandwidth, and its negative effects are more important at higher frequencies where the tag response has a lower amplitude.

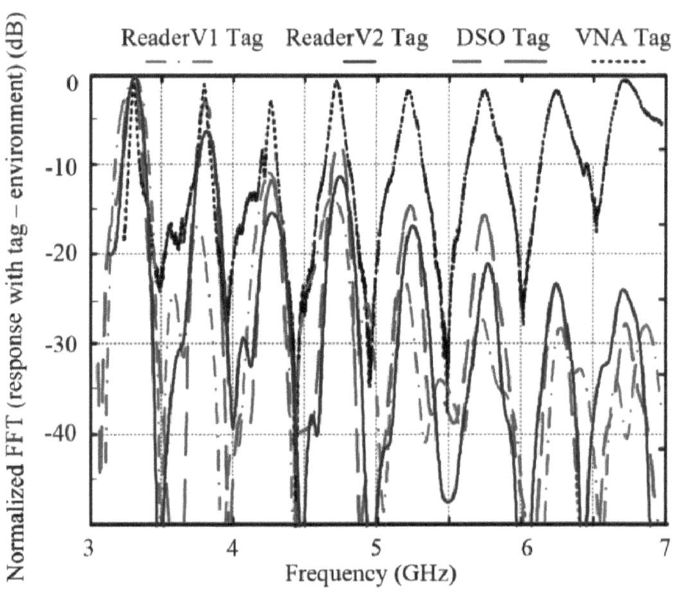


**Figure 4.17.** *A chipless tag acquisition with different reading systems, which are based on SFCW and IR-UWB with different performance in terms of jitter. The VNA N5222A from Keysight emulates an SFCW-based reader, while the first reader version (Chapter 3) and the second version show jitter effects. For a color version of this figure, see www.iste.co.uk/garbati/chipless.zip*

These experiments have been performed without taking into account the UWB regulations, which, as discussed in Chapter 2, affect frequency-based readers more than IR-UWB.

## 4.9. Conclusion

An optimized UWB chipless RFID reader version has been introduced. It was designed to reduce its sampling noise, with improvements made at RF hardware and software levels. The design of the Xilinx embedded systems was carefully optimized, not only to reduce sampling noise, but also to massively reduce the reading time. The reader is fully configurable in terms of acquisition resolution (equivalent sampling rate), frame duration and position. The reading time is in the order of a few ms, depending on the chosen parameters. Reader sensitivity was about −60 dB; this value is more limited by the ADC resolution, which can be increased using a different model. The retail cost of reader components is estimated to be 2,306 € after tax. A reduction of about 20/30% is feasible in industrial production. Reader performances and characteristics have been summarized and shown in Table 4.1.

To reduce the cost of reader components, a pulse generator has been designed, realized and characterized. It is presented in the following chapter, and potentially, it could have a retail component cost of about 200 €, which is five times lower than the cost of the actual reader pulse generator. As expected, a more integrated reader version can be designed, where the RF front-end, the new pulse generator and the used components of the SP601 Xilinx evaluation board are on the same board. This should free up the estimated cost of the Xilinx evaluation board, which is about 420 €, while the employed components have a total cost of about 50 €. The use of a higher performance substrate than FR-4 at higher frequencies will also increase the reader's capability. The components' total retail cost of 1,000 € is feasible. Finally, the design of a low-cost UWB antenna is also presented in the following chapter. This may be used to propose a complete low-cost IR-UWB chipless RFID reader solution.

# 5

# UWB Pulse Generator and Antenna Design

## 5.1. Introduction

In this chapter, the design, realization and characterization of a UWB pulse generator and a planar UWB antenna are described. The pulse generator was designed to have a low jitter and low cost, and high-performance discrete digital commercial off-the-shelf (COTS) components were used for its realization. The designed UWB antenna was a planar Vivaldi in the bandwidth of 3–10 GHz. A pulse generator with Ansys Electromagnetics suite 17.0 and an antenna with CST Studio Suite 2014 were designed and simulated. This chapter is organized as follows:

Section 5.2 presents the designed pulse generator with simulation and measurement result.

Section 5.3 shows the measurement of frequency-coded tags with the designed pulse generator. The pulse generator is also used as the emission stage of the optimized reader prototype presented in Chapter 4 to read the same tags.

Section 5.4 presents the designed UWB antenna.

Section 5.5 draws the conclusions of this chapter.

## 5.2. UWB pulse generator design

The optimized reader presented in Chapter 4 uses a commercial pulse generator in the emission stage. At present, HL9200 from HyperLabs is obsolete and substituted by the newer version HL9201. It shows similar performance to the predecessor, with a retail cost of $1.295. In this section, we present a home-designed fully tunable baseband UWB pulse generator realized with COTS components, which is intended to be less expensive and with low jitter. It is made up of high-speed logic gates and comparators and complies with the current mode logic (CML) digital standard. The pulse-emitted power and bandwidth are tunable, showing high flexibility. This advantage simplifies the design of the downstream shaping network to be in compliance with international UWB regulations. A differential output with a very low jitter is demonstrated.

### 5.2.1. *Introduction to UWB pulse generator design*

UWB technology has various applications ranging from radar for tracking and localization to short-pulse communication [FON 04]. A UWB radar transmits a series of UWB pulses in the direction of a target and then measures the backscattered signals. These sources of information are used to recover the target position and shape. For short-pulse communication applications, time or frequency-division multiplexing algorithms are used. Both are real-time implementations [FON 04] and usually adopt diode-based receivers, such as a frequency converter, to recover the data in baseband. Owing to the receiver characteristics, a high reproducibility of the transmitting UWB pulse is not required. Therefore, a time-gated oscillator is generally employed as a UWB pulse generator. Besides pulse position modulation (PPM), it is possible to use pulse amplitude modulation (PAM), where the pulse generator output amplitude is attenuated depending on the transmitting binary code [MOO 12].

In order to reduce the fabrication costs, most UWB radars are based on an equivalent-time algorithm that allows an effective time resolution to be achieved in the order of a few picoseconds [WAN 12]. Using this approach, the reproducibility of the UWB transmitting pulse is a key parameter, and therefore the pulse generator cannot be a simple time-gated oscillator. A baseband UWB pulse generator, hereafter B-UWB-PG, needs to be employed. It is capable of generating a Gaussian pulse of different orders which can be derived in the time domain, with a dedicated downstream circuit, depending on the project requirement. This circuit can be formed of only the antenna as a result of its time derivative property [SCH 05b].

According to the application, B-UWB-PGs differ in characteristics [DEA 08]. Most are designed with step recovery diodes (SRDs) and bipolar transistors driven in avalanche mode to generate a Gaussian baseband pulse [XIE 06]. Owing to the voltage and frequency output characteristics, the SRD solution is commonly used in various applications, ranging from UWB communications to system characterization [OH 11, LI 12]. The pulse is sub-nanosecond with a high slew rate, i.e. a low pulse rise and fall time. The B-UWB-PG is composed of two different circuits in series, a driver and a pulser [HEW 84]. The driver can generate a pulse waveform with a duration of several hundred nanoseconds. The pulser exploits the junction impedance transition of the diode to generate the UWB pulse. This solution can provide a pulse repetition frequency (PRF) in the order of a few megahertz (MHz). A bipolar transistor driven in avalanche mode solution [DEA 07] provides a high-voltage output pulse at the expense of a limited PRF and the requirement of a high-voltage supply (about 200 V). It is ideal for long-distance radar utilization.

The reproducibility of a step recovery diode (SDR)-based pulse generator output was studied by Kheirdoost, Moradi, Elkholy *et al.* [KHE 14] with its jitter modeling. It shows how defects in pulse reproducibility negatively affect equivalent-time sampler systems rather than the real-time ones. The random jitter of the complete device (which represents the statistical variation of its propagation delay) increases rapidly, mostly due to the shot noise of the diode. The statistical jitter standard deviation can be in the order of tens of picoseconds [KHE 14]. The UWB chipless RFID systems are suitable for equivalent-time receivers thanks to their stationary nature. Thus, the pulse generator is expected to have low jitter as well as low cost [GAR 15, PER 14], thereby ensuring an affordable and reliable reader.

It is also possible to design a B-UWB-PG with logic gates, in which wideband glitches are created. Almost all of them are integrated solutions [REJ 13, LEE 14] based on sharped trigger input signals, with limited flexibility in terms of pulse width and amplitude, with single-ended output.

A UWB pulse generator realized with COTS components was proposed in Schwoerer, Miscopein, Uguen *et al.* [SCH 05a]. Its architecture is based on an inverter and one OR gate, its maximum PRF is about 100 MHz, the pulse width is fixed at 500 ps and there is no flexibility in terms of pulse width and output amplitude. Therefore, the design of the pulse shaping network, to be in compliance with international regulations, may be cumbersome. The regulations limit the UWB power emitted according to the FCC and the ETSI [ETS 17, FCC 16]. Finally, owing to its high pulse width, the signal energy at higher frequency can be extremely low for applications such as UWB chipless RFID.

To overcome these limitations, a B-UWB-PG made up of high-frequency discrete logic is proposed. Indeed, nowadays, commercial discrete digital devices with clock frequencies over 10 GHz are available for applications such as RF test equipment and serial data transmission. The proposed solution does not require any input trigger with a sharp rising edge. This feature is a relevant advantage in practical applications, because it can be integrated with low-speed transistor–transistor logic (TTL) and complementary metal-oxide semiconductor (CMOS) technologies-based system. It also presents a tunable pulse width and amplitude feature, which may be exploited to reduce the complexity of the downstream circuit design. The circuit was intentionally designed with COTS components to be easily reproduced and customized around diverse project requirements. This unloads for an eventual ASIC design, which is generally expensive and time-consuming.

### 5.2.2. *B-UWB-PG architecture*

Figure 5.1 shows a schematic diagram of the proposed device. It is mainly composed of two high-performance comparators, *A* and *B*, a variable delay generator and an *AND* gate. According to the CML standard, the connections between these devices are balanced. The connection between the comparator *A* and the *IN 1* input of the *AND* gate has a tunable length by soldering 0402 0 Ω surface-mount resistors. It is possible to add three different additional transmission lengths, denoted as *A1-A2-A3*, where $A2 = 2A1$ and $A3 = 2A2$, with $A1 = 16$ mm. The positive input of *A* and the negative input of *B* are connected with the trigger input, and the other two inputs are connected with the positive threshold voltage *VT*.

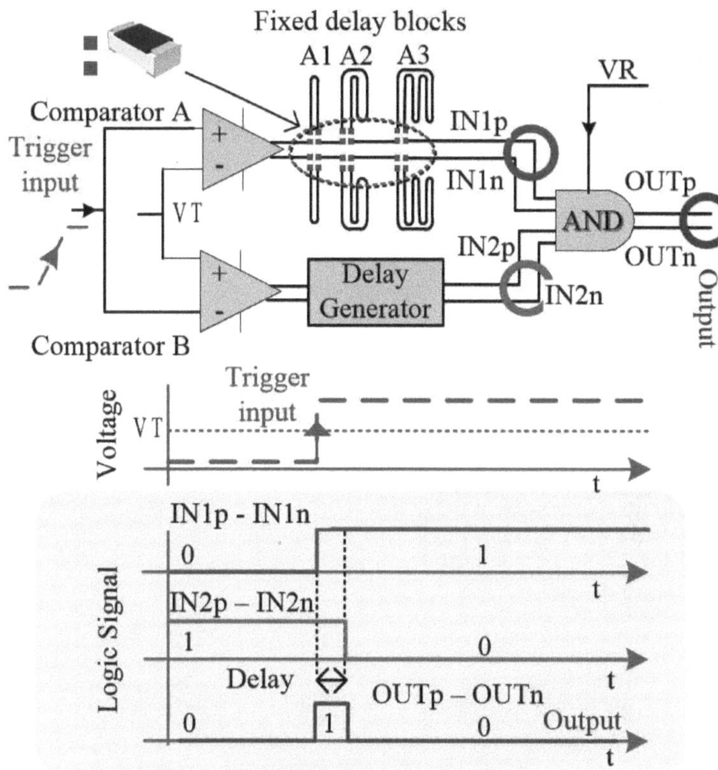


**Figure 5.1.** *Schematic of the B-UWB-PG principle. After a trigger event (rising edge), the comparator A switches from $0_2$ to $1_2$, and the comparator B switches from $1_2$ to $0_2$. Tuning the delay generator, the AND gate has its two inputs with the same logical value 1 for a short period of time. It creates a glitch, i.e. a baseband UWB pulse. For a color version of this figure, see www.iste.co.uk/garbati/chipless.zip*

When the trigger input voltage crosses the threshold, the two comparators simultaneously switch in opposite digital directions. If the trigger input is at $0_2$, i.e. lower than the threshold, the output of *A* is $0_2$ and that of *B* is $1_2$. Therefore, the output of the *AND* gate is $0_2$. When the trigger crosses *VT* during its rising edge, the output of *A* will go to $1_2$ and that of *B* will go to $0_2$. However, the inputs of the *AND* gate do not switch simultaneously. The time needed for the signals to travel from the output pins of the corresponding comparator to the input pins of the *AND* gate for IN1 also depends on the length of the fixed delay blocks *A1-A2-A3* when they are employed. While for *IN2* it depends on the delay generator. The modification of these parameters results in the same logical value $1_2$ for the two inputs of the AND gate for a tunable and short period of time. It produces a glitch at the gate output, which is the generated pulse. The *AND* gate presents a pin called *VR* that can continuously modulate the output amplitude. Its absolute amplitude variation was experimentally found within 0.6 V, with the minimum at 0.03 V for *VR* = −1.2 V and the maximum for *VR* = 0.4 V.

The evaluation board of the proposed pulse generator is shown in Figure 5.2. It is a four-layer board with dimensions of 8.5 cm × 7 cm, manufactured using a low-cost FR-4 substrate.

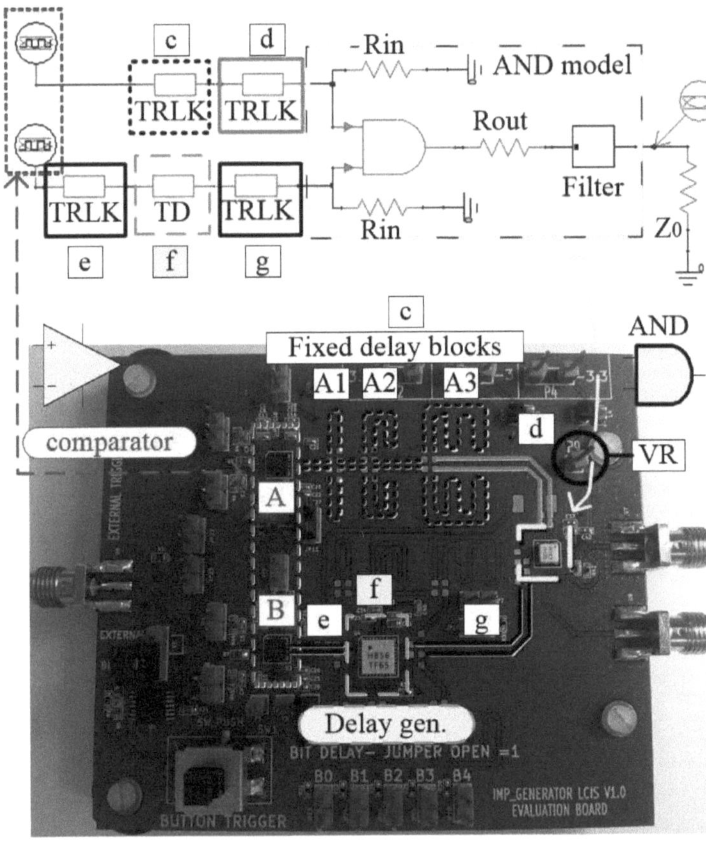


**Figure 5.2.** *B-UWB-PG evaluation board. This four-layer FR-4 PCB presents three different external power voltage values: 3 V, −3 V and −3.3 V. Through a jumper, the trigger could be external or entrusted on a push button. The delay of the delay generator is imposed by five jumpers. The output is balanced, and once DC coupled, it is capable of generating two differential output pulses. The pin VR is used to modulate the output amplitude. For a color version of this figure, see www.iste.co.uk/garbati/chipless.zip*

The selected comparator was HMC675LP3E from Analog Devices, which can output a signal with rise and fall times of 27 and 18 ps, respectively. It presents a deterministic jitter of 2 ps, a random jitter standard deviation of only 0.2 ps and a balance output compatible with the CML digital standard. The delay generator is the Analog Devices HMC856LC5 with rise and fall times of 20 and 18 ps, respectively. It is compatible with the CML standard and presents a balance input and output. Its propagation delay is 255 ps with an additional digitally controlled delay ranging from 0 to 93 ps, with 3 ps resolution. The AND gate is the Analog Devices HMC722LCRC, with output rise and fall times of 19 and 18 ps, respectively. It has a deterministic jitter of 2 ps and a random jitter of 0.2 ps. All the components were chosen for their performance in terms of jitter and output rise and fall times (high slew rate).

The balanced interconnections are realized with differential parallel-grounded coplanar waveguide (GCPW) transmission lines. The glitch signal, i.e. the output of B-UWB-PG, must have a high slew rate and pulse width in the order of tens of picoseconds so as to ensure a large −10 dB bandwidth over UWB (3.1–10.6 GHz). The retail cost of the main pulse generator components after tax is given in Table 5.1. The total cost with the delay generator is about 381 €. The delay generator is the most expensive component, i.e. 184 €. If the fine variable output pulse-width-tuning characteristic is not required, its use is not essential. Thus, a low-cost pulse generator with a retail cost of the main components of 200 € after tax may be designed.

| **Pulse generator's retail cost after tax** | |
|---|---|
| **Component** | **Retail cost after tax (2017 (Digikey))** |
| **Comparator HMC675LP3E** | 35€ × 2 |
| **Delay generator (HMC856LC5)** | 184 € |
| **AND gate (HMC722LCRC)** | 127 € |
| **Total cost with delay generator** | 381 € |
| **Total cost without delay generator** | 197 € |

**Table 5.1.** *Components' retail cost of the presenting pulse generator after tax*

With respect to the blocks A1-A2-A3 for the connection between comparator *A* and *AND* gate, as well as with the delay generator, it is possible to obtain a pulse with a full width at half-maximum (FWHM) (pulse width) ranging from 325 to 60 ps with a resolution as low as 3 ps. The delays added with the blocks *A1-A2-A3* are, respectively, about 100, 200 and 400 ps. The PRF is only limited by the pulse width and the input bandwidth of the comparator. The latter has an input bandwidth of 10 GHz, while the maximum pulse width is 325 ps. If the input trigger has a duty cycle of 50%, then the PRF is approximately

$$PRF = min(10\,GHz,\ \frac{1}{2PW})), \qquad [5.1]$$

where *PW* is the pulse width. For *PW* = 325 ps, *PRF* = 1.53 GHz, while for *PW* = 60 ps, PRF = 8.3 GHz. The limit expressed in equation [5.1] can be increased by using a trigger signal with a higher duty cycle.

### 5.2.3. *B-UWB-PG simulation model*

#### 5.2.3.1. *Block description*

Figure 5.1 shows the simulation model of the pulse generator prototype realized using Ansys Electromagnetics suite 17.0. The output of the two high-performance comparators was emulated using two eye sources. They are capable of transmitting two series of pulses, with rise and fall times congruent with the comparator specification. Furthermore, a source of jitter with Gaussian PDF was added. The chosen standard deviation of 2 ps is higher than the value reported in the data sheet of the comparators. This choice can be justified for real applications, where many factors, such as environmental conditions, impedance mismatching and crosstalk, may increase the effective jitter of the electronic devices [MIT 09].

The transmission lines between the comparators output and the *AND* gate input were modeled with *TRLK* blocks. They are transmission lines where all the fundamental parameters are assumed not dispersive. A variable accounts for the different lengths of blocks *A1-A2-A3*, and is implemented with the *TRLK*, denoted by *c*. The *TRLK* of 53.04 mm, denoted by *d*, takes into account the length of the lines, which connects the comparator A and the corresponding *AND* gate input after the blocks *A1-A2-A3*. The connection between the output of *B* and the input of the delay generator is represented with the *TRLK* block, denoted by *e*, which has a length of 12.23 mm. The *TRLK* block, denoted by *g*, is 35.5 mm for the connection between the output of delay generator and the other input of the *AND* gate. Another variable accounts for the delay generator propagation and tunable delay, and is implemented using a *TD* block, which is an ideal transmission line defined by its impedance characteristics and time delay.

The *AND* gate was modeled using a simple ideal threshold model written in Verilog-AMS, where AMS stands for analog and mixed signal extension, three 50 Ω resistors for input and output impedances and finally a fifth-order Gaussian low-pass filter with a bandwidth of 10.84 GHz. This value was obtained using the well-known relationship between the filter output rise time at an ideal step input and the filter bandwidth reported in equation [4.7] of Chapter 4 [DEL 98]. The output rise time was that of the *AND* gate data sheet multiplied by an empirical factor of 1.6. This factor is used to estimate the output 10%–90% rise-time from the 20%–80% rise-time shown in the AND gate data sheet. In Figure 5.2, *Z0* represents an ideal 50 Ω load. The AND gate output jitter was added in post-processing using Matlab®, with Gaussian PDF with a 2 ps standard deviation from the Ansys transient simulation result. For simplicity, all the connections between the different components were single rather than balanced.

Figure 5.3 shows the eye diagram of the simulation model output tuning the variable that models the delay generator. The color bar represents the vertical histogram results of the eye diagram in logarithmic scale. It was realized over 300 pulse realizations with Matlab®. The *TRLK*, denoted by *c*, was fixed at 32 mm (*A2*), while the variable representing the delay generator varied in the following range of discrete values: 279 ps (255 ps + 24 ps), 285 ps (255 ps + 30 ps), 300 ps (255 ps + 45 ps), 315 ps (255 ps + 60 ps) and 348 ps (255 ps + 93 ps). The smaller pulse width of Figure 5.3 of about 30 ps corresponds to the minimum delay imposed of 279 ps, and the larger pulse width of about 100 ps corresponds to the maximum delay of 348 ps. In the simulation model, the effect of the output amplitude variation using the pin *VR* was implemented, but not reported because it varies the output amplitude only linearly without affecting the system performance. It is evident from Figure 5.3 that for the signals with a pulse width less than 50 ps, the amplitude also decreases.

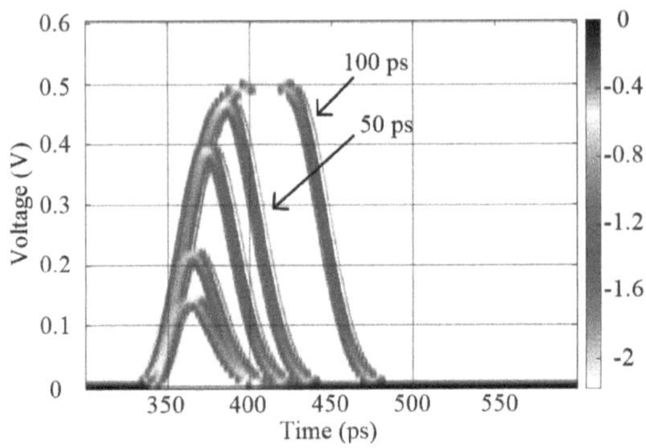


**Figure 5.3.** *Eye diagram of the simulation model presented in Figure 5.2 obtained by tuning the variable that models the delay generator. It was realized over 300 pulse realizations, where the jitter of the AND gate output was added from the Ansys transient simulation results, during post-processing with Matlab®. For a color version of this figure, see www.iste.co.uk/garbati/chipless.zip*

### 5.2.3.2. *Noise rejection*

The use of high-performance comparators in terms of low rise and fall times is indispensable for practical applications. This is because the digital logic devices employed do not have hysteresis capability. Therefore, their input–output characteristic is a static nonlinear function, where the dynamic effect is only due to parasitic components. The proposed prototype is designed to be integrated with onset UWB systems, where it may share the same PCB with other blocks of circuit. The proposing prototype is composed of transmission lines with length in the order of a few centimeters between the comparator output and the AND gate input. Thus, it may suffer from crosstalk effects. In this scope, different simulations have been performed making use of the model shown in Figure 5.2.

The crosstalk effects have been emulated with two sinusoid voltage generators, Vup and Vdown. They were inserted at the output of the two eye sources that simulate the output of the two comparators. The amplitudes of the Vup and Vdown were fixed at 40 mV, while the frequency was varied from a few GHz to tens of GHz. The B-UWB-PG was simulated with a fixed configuration, corresponding to a pulse width of approximately 90 ps, and the simulation was carried out acting on the rise and fall times of the eye sources. First, they were settled at 27 and 18 ps, corresponding to the characteristics of the two actual comparators employed. The corresponding simulation results are shown in Figure 5.4(a). Then, the rise and fall times of the eye sources were settled at 1 ns, which can correspond to the rise and fall times of TTL and CMOS devices. The corresponding simulation results are shown in Figure 5.4(b). In the case of the lower rise and fall times (27/18 ps) (Figure 5.4(a)), a high rejection at crosstalk effect is demonstrated. In the case of higher rise and fall times (1 ns), the B-UWB-PG output has an increased jitter due to the crosstalk for the lower frequency values of Vup and Vdown. On the contrary, for higher frequency values, the B-UWB-PG output distortion is prominent. Thus, the simulation results validate the use of high-performance comparators with low output rise and fall times ((27/18 ps)).

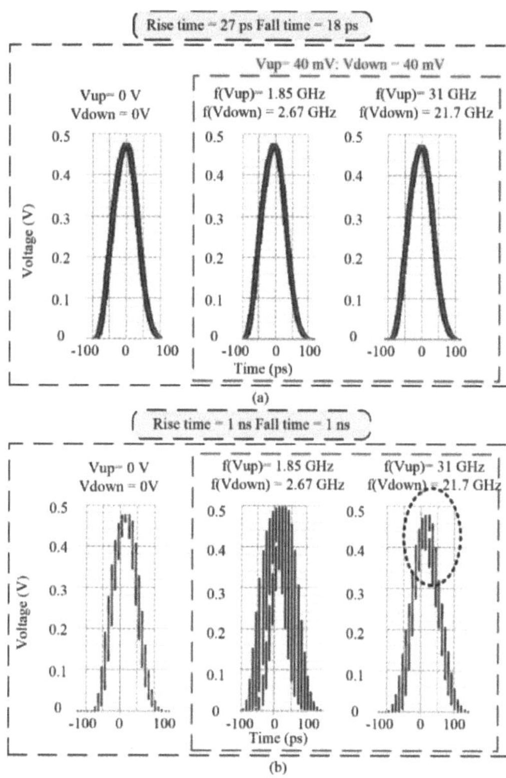


**Figure 5.4.** *Simulation of the crosstalk effect on the proposed B-UWB-PG. a) Simulated eye diagrams of the B-UWB-PG output in the case of the eye sources' rise and fall times of 27 and 18 ps, respectively. They correspond to the output rise and fall times of the actual comparators used for the design of the B-UWB-PG. The B-UWB-PG was simulated with a fixed configuration, corresponding to a pulse width of approximately 90 ps. b) Simulated eye diagrams of the B-UWB-PG output in the case of the eye sources' rise and fall times of 1 ns, which can correspond to rise and fall times of either a TTL or a CMOS device. For a color version of this figure, see www.iste.co.uk/garbati/chipless.zip*

### 5.2.4. *Measurement results*

#### 5.2.4.1. *B-UWB-PG output characterization*

As the B-UWB-PG presents a balanced output, two pulses with inverted polarity can be generated. According to the CML digital standard, the DC component of the two output pins differs. It is 0 V for the negative single output and –0.4 V for the positive output when no pulse is created. However, it is possible to force the DC component of both pins at 0 V once the AC is coupled to ground.

Measurements of the positive and negative output voltages of the proposed pulse generator, with different FWHM at VR = 0 V, are shown in Figures 5.5(a) and (b), respectively. The pulse amplitude varies with pulse width from the FWHM about 90 ps, while it was 50 ps from the simulation results shown in Figure 5.3. This could be due to the filtering effect at a higher frequency of the realized PCB.

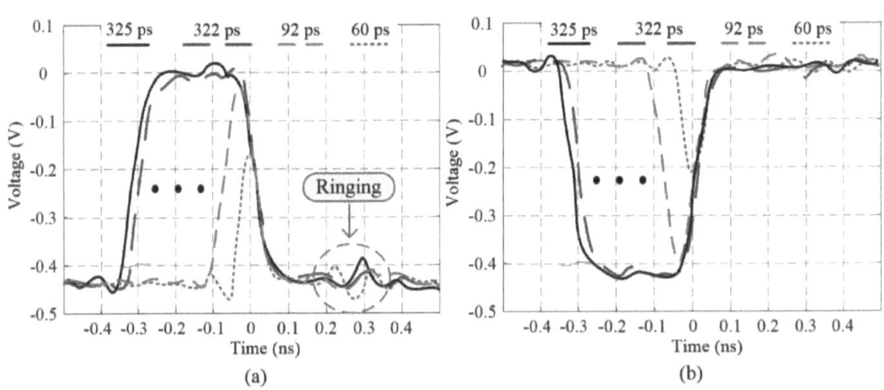


**Figure 5.5.** *Single output of the B-UWB-PG evaluation board measured for different delay generator configurations and transmission line lengths (A1-A2-A3) with VR = 0 V. The measured a) positive and b) negative balanced outputs of the AND gate, both of which are referred to ground. For a color version of this figure, see www.iste.co.uk/garbati/chipless.zip*

The corresponding PSD over MHz of the positive output voltage (Figure 5.5(a)) is shown in Figure 5.6. It was estimated with Matlab® supposing a PRF of 1 MHz, which is that of the optimized reader described in Chapter 4. Its absolute values can be increased to a maximum of 2.5 dB varying *VR* from 0 to the maximum value of 0.4 V. As expected, the maximum response is at 0 Hz, independent of the pulse width [GHA 07]. The –10 dB bandwidth for the 60 ps is about 10 GHz. To modulate the position of the peaks and valleys at the frequency range, it is essential to derive the response of the pulse generator in the time domain.

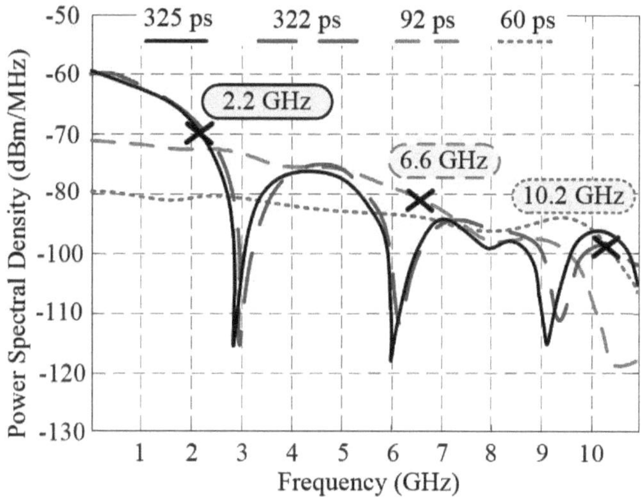


**Figure 5.6.** *Calculated PSD/MHz with a PRF of 1 MHz of the signals plotted in Figure 5.5(a) with a PRF of 1 MHz. For a color version of this figure, see www.iste.co.uk/garbati/chipless.zip*

From Figure 5.5, the shape of the pulses can be approximated with a rectangular function for larger FWHM, while Gaussian for smaller FWHM. The module of the spectrum of a rectangle of width $\Delta T$ and amplitude $A$ is

$$|G(f)| = |A\Delta T sinc(\pi f \Delta T)|, \qquad [5.2]$$

and its –10 dB bandwidth is approximately

$$f_{-10dB} = \frac{0.73}{\Delta T}. \qquad [5.3]$$

It is evident from Figure 5.5(a) that the larger pulse FWHM is 325 ps, and according to equation [5.3], it should have a –10 dB bandwidth of 2.25 GHz, rather than the smaller measured value of 2.2 GHz. The error with respect to 2.25 GHz is only 2.2%, which validates the rectangular pulse assumption. The narrower pulse shapes can be approximated with Gaussian pulses with the expression

$$V(t) = \frac{A}{\sqrt{2\pi\sigma^2}} e^{(-\frac{t^2}{2\sigma^2})}. \quad [5.4]$$

where $A$ is the maximum of the pulse and σ is the shape factor (standard deviation). For simplicity, we assume that the curve is centered in 0 s. The module of its spectrum is

$$|G(f)| = \left|Ae^{(-\frac{(2\pi f\sigma)^2}{2})}\right|, \quad [5.5]$$

and its –10 dB bandwidth is

$$f_{-10dB} = \frac{\sqrt{2.303}}{2\pi\sigma}. \quad [5.6]$$

For Gaussian pulse, the relationship between FWHM and shape factor is

$$FWHM = 2\sqrt{2ln2\sigma} \approx 2.3548\sigma. \quad [5.7]$$

Therefore, combining equations [5.6] and [5.7], we obtain

$$f_{-10dB} = \frac{0.5687}{FWHM}. \quad [5.8]$$

From Figure 5.5(a), the minimum FWHM is about 60 ps. Using equation [5.8], its –10 dB bandwidth should be about 9.48 GHz. The measured value differs by only 5.5% and is about 10.2 GHz. It can be demonstrated that this error is due to the ringing effect shown in Figure 5.5(a). For the case of FWHM = 92 ps, from equation [5.8], the –10 dB bandwidth should be 6.18 GHz, but it is 6 GHz. This small difference is reduced with respect to the previous case due to the near-absence of pulse ringing [MIT 09].

Figure 5.7 shows the measurement results for the negative (gray dashed lines) and positive (black lines) outputs of the proposed B-UWB-PG, for two FWHM configurations of 235 and 126 ps, respectively, varying linearly the voltage applied to the pin *VR* from 0.4 and –1.2 V. Figure 5.7 shows a dynamic variation for each port from 0.6 to 0.03 V.

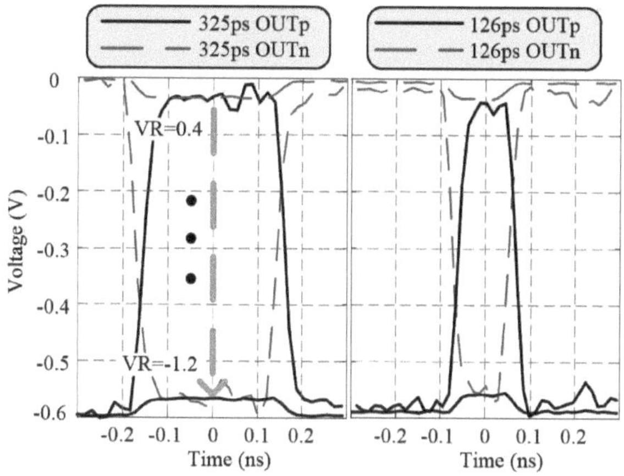


**Figure 5.7.** *Measurement results for the negative (gray dashed lines) and positive (black lines) outputs of the proposed B-UWB-PG for two FWHM configurations of 235 and 126 ps, respectively, varying linearly the voltage applied to the pin VR from 0.4 to −1.2 V*

### 5.2.4.2. *B-UWB-PG jitter characterization*

The jitter of the pulse generator was measured using the bench shown in Figure 5.8(a). It is composed of DSO Agilent DSO91204A (with 12 GHz bandwidth and 40 GSa/s sampling rate), the Agilent 1134A high-performance probe and an arbitrary function generator (AFG) AFG3102C from Tektronix.

The AFG is used to trigger the pulse generator using a periodic square signal of 1 MHz (PRF). The probe is placed at the input of the comparator A and used as a trigger for the DSO, which is connected to the pulse generator output. The eye diagram for a pulse with an FWHM of approximately 60 ps over 1,000 pulses is shown in Figure 5.8(b). The pulse-width duration presents a standard deviation of 1.6 ps and a peak-to-peak of 8.33 ps. The standard deviations of fall and rise times (20–80%) are 1 and 0.8 ps, respectively. These values are in accordance with the jitter specifications of the high-performance components employed. In order to compare with a commercial solution, the jitter standard deviation of the commercial solution HL9200 from HyperLabs, employed as the transmitting stage of the readers described in Chapters 3 and 4, is 3 ps from the data sheet. The presenting solution shows lower jitter results, and with a high-performance RF board substrate, it would further decrease.

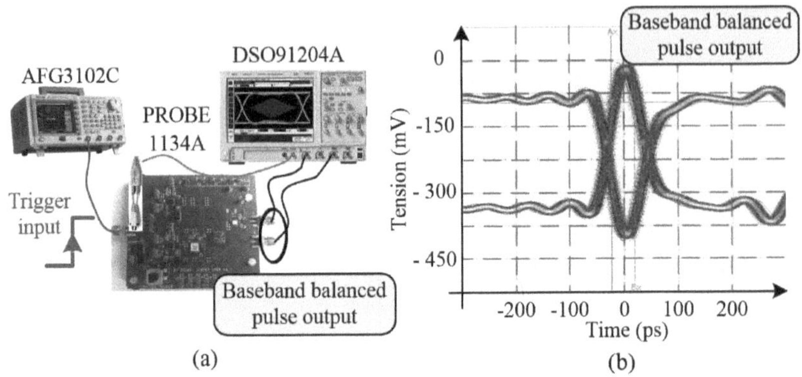


**Figure 5.8.** *Deterministic period jitter measurement with an eye diagram using the 12 GHz Agilent DSO91204A and the 7 GHz Agilent 1134A probe as trigger reference for the oscilloscope. a) Test bench. b) Measurement results. For a color version of this figure, see www.iste.co.uk/garbati/chipless.zip*

### 5.2.5. *Output stage considerations*

#### 5.2.5.1. *Output power enhancement*

The maximum PSD/MHz of each single B-UWB-PG output, with VR = 0 V, ranges from –59 to –80 dBm/MHz (see Figure 5.6) for pulse widths of 325 and 60 ps, respectively. This low level of power is due to the low duty cycle used, i.e. low PRF. Increasing the PRF will result in a higher PSD/MHz. To increase the output power, the low-cost Analog Devices HMC788 gain block amplifier was used as the final stage of the proposed B-UWB-PG. It is characterized by an operative band from DC to 10 GHz, a gain of 14 dB, an input compression point of 6 dBm and a noise figure (NF) of 7 dB. An evaluation board for the amplifier was fabricated on the same FR-4 substrate of the pulse generator. The measurement test bench is shown in Figure 5.9, in which the amplifier is connected in cascade with one output of the B-UWB-PG. The measured output voltage is shown in Figure 5.10(a), while its estimated PSD/MHz is shown in Figure 5.10(b). The proposed solution shows an output amplitude ranging from 1.1 to 2.3 V with VR = 0 V. Comparing the same results without the output amplifier, the –10 dB level of the pulses decreases. It was mainly caused by the negative decay of the gain of the amplifier with frequency.

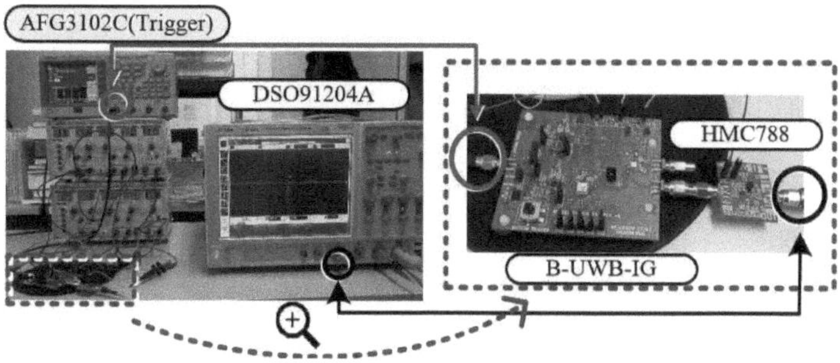


**Figure 5.9.** *Measurement setup of a single output of the B-UWB-PG amplified with the evaluation board for the HMC788 gain block amplifier. For a color version of this figure, see www.iste.co.uk/garbati/chipless.zip*

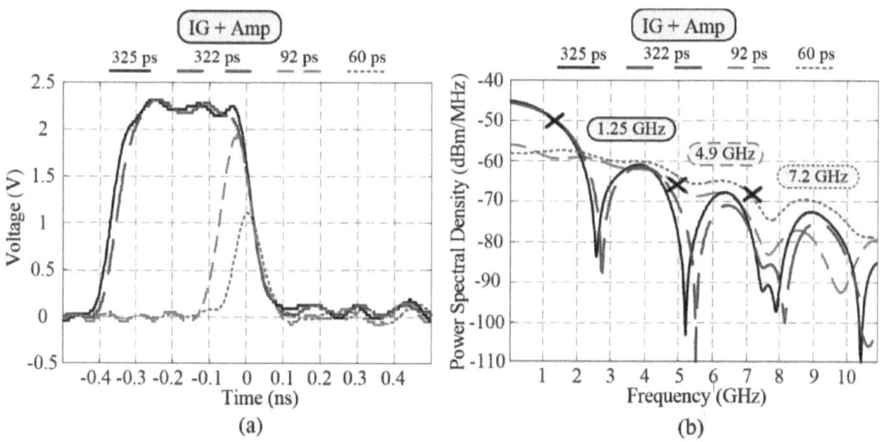


**Figure 5.10.** *a) Measurement results of the B-UWB-PG with the gain block amplifier (see Figure 5.9) with time. b) Calculated PSD/MHz with a PRF of 1 MHz. For a color version of this figure, see www.iste.co.uk/garbati/chipless.zip*

### 5.2.5.2. *UWB shaping network*

In order to comply with FCC regulations in terms of power mask, the output of the proposed B-UWB-PG must be shaped [FCC 16]. According to the regulations, the reported value of PSD/MHz is a measure of the EIRP; therefore, the characteristics of the transmitting antenna should be taken into account. Outside of the 3.1–10.6 GHz band, the allowed PSD is strictly limited, especially between 1 and 2 GHz. The UWB antenna itself can be used to shape the pulse [TAM 12] or the antenna in combination with a band pass filter [MEN 05].

Typically, a monocycle pulse, which is the result of a first derivation of a Gaussian pulse, is used at the transmitting port of commercial radars, such as the Novelda NVA-R640 kit. The corresponding maximum PSD is not located at zero frequency. By combining the balanced outputs of the proposed B-UWB-PG pulse generator, it is possible to generate a monocycle Gaussian pulse. Therefore, the B-UWB-PG can be used with most commercial UWB antennas while respecting the regulations.

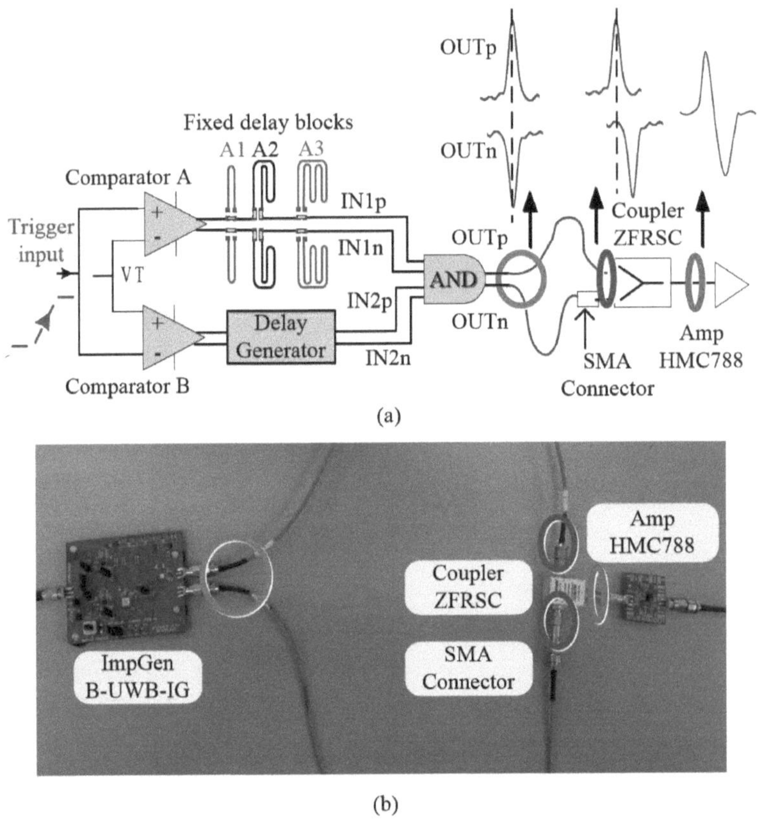


**Figure 5.11.** *a) Schematic and b) photograph of the monocycle pulse generator obtained from the combination of the balanced outputs of the proposed B-UWB-PG, using a power combiner and a short delay line. For a color version of this figure, see www.iste.co.uk/garbati/chipless.zip*

Figure 5.11(a) shows a schematic diagram of the combination process. The monocycle is obtained by carrying out summation of two Gaussian pulses, one positive and one negative, with the latter being slightly delayed in time. The wideband combiner Mini-Circuits ZFRSC-123 was used to perform the summation. The time shifting between the two pulses was carried out using two cables with different overall lengths, which, in this case, is a 21 mm coaxial connector that adds a small delay to the bottom chain (see Figure 5.11(b)). In order to compensate the insertion loss of the combiner, which is about 10 dB, the same HMC788 amplifier was employed as a final amplification stage. The results are shown in blue dashed lines, compared with the NVA-R640 commercial radar pulse output in black line in Figure 5.12. The PRF was increased to 48 MHz, which corresponds to that one of the Novelda NVA-R640 kit. Both responses are similar, showing that the B-UWB-PG can be easily tuned to provide a monocycle pulse. In the case of our prototype, the pin VR was at 0 V, and therefore, the output amplitude could be yet increased using the maximum value of 0.4 V. The higher power level of Figure 5.12, compared with Figure 5.10(b), is due to the higher PRF.

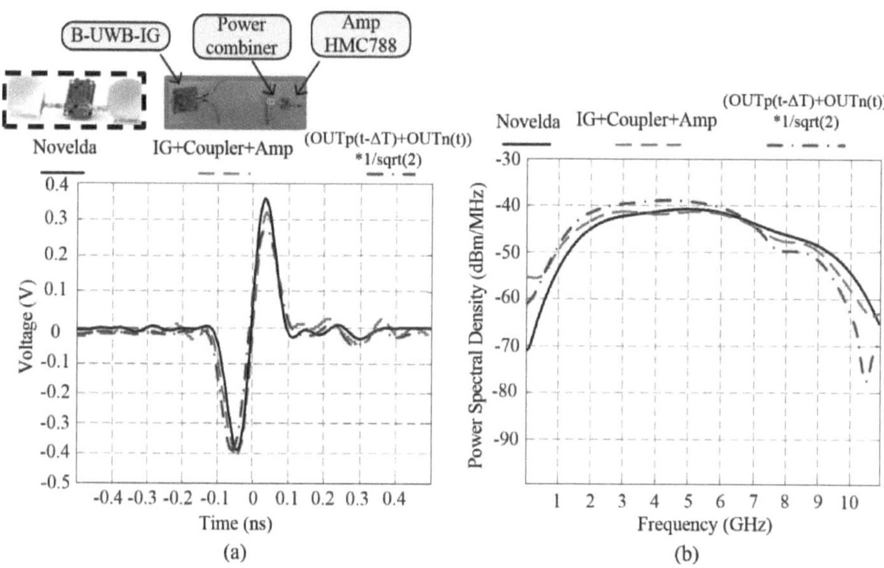


**Figure 5.12.** *Dashed blue line: measurement of the proposed pulse generator with the wideband combiner ZFRSC-123 from Mini-Circuit and HMC788 amplifier (see Figure 5.11) with VR = 0 V. Red dash-dotted line: distinct single output measurement from the B-UWB-PG, called OUTP and OUTN, from Figure 5.11(a), and summed during post-processing with a delay of 14 ps executed with VR = 0.4 V. Black line: Novelda NVA-R640 development kit pulse output. a) Signal in the time domain. b) Calculated signal in the frequency domain (PSD/MHz), where the PRF was that of the Novelda NVA-R640, corresponding to 48 MHz. For a color version of this figure, see www.iste.co.uk/garbati/chipless.zip*

The wideband combiner Mini-Circuits ZFRSC-123 has a high insertion loss due to its large band, from DC to 12 GHz. With a lower insertion loss combiner, and using 0.4 V for *VR*, it would be possible to eliminate the final amplifier. Figure 5.12 also shows, in red dash-dotted line, the distinct single output measurement from the B-UWB-PG, called OUTP and OUTN, from Figure 5.11(a), executed with VR = 0.4 V. They were summed in post-processing in the time domain with a delay of 14 ps, and finally attenuated to 3 dB to emulate an ideal power combiner. The signal in time and frequency presents almost the same trend as the case with the amplifier (blue dashed line).

Finally, the same configuration of Figure 5.11 was used, where the combiner was substituted with the Mini-Circuits ZFSC-2-9G+ band pass combiner. The filtering effect of the new combiner is clearly visible in Figure 5.13(a). The estimated PSD/MHz is shown in Figure 5.13(b), in which the gain of the Satimo QH2000 antenna was taken into account for estimating the EIRP. The PRF was 1 MHz, and the FCC average power mask for indoor applications is shown in the figure. As observed, the final emitted pulse is in full compliance with the regulations. Moreover, the output power can also be increased.

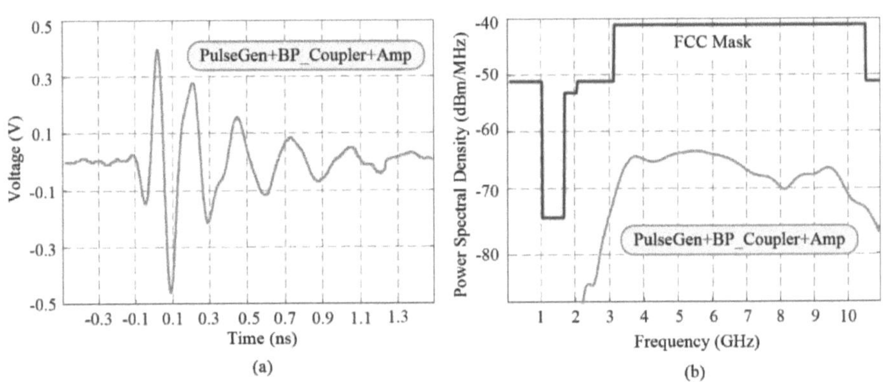


**Figure 5.13.** *a) Measurement in the time domain of the setup of Figure 5.11(b) with the ZFSC-2-9G+ band pass combiner. b) Calculated PSD/MHz with a PRF of 1 MHz taking into account the gain of the Satimo QH2000 antenna, and FCC average power mask for UWB indoor applications*

### 5.2.6. *Conclusion*

This section has presented a potentially low-jitter fully tunable baseband pulse generator for real- and equivalent-time UWB applications. COTS components and low-cost FR-4 substrate have been used. The generator is based on high-frequency logic gates and two comparators. It can be triggered by a low-frequency TTL-CMOS digital signal. A delay generator and variable-length transmission lines allow the tuning of the pulse width from 325 to 60 ps. A pin on the *AND* gate allows for amplitude modulation. The measured random period jitter over 1,000 cycles is on the order of 2 ps. The generator outputs two pulses that can be combined to create a commonly used monocycle pulse. The proposed solution is flexible and allows a shaping network to be designed with COTS components to comply with the regulations. The use of COTS components is a big advantage compared to an integrated solution because their implementation, modification and redesign are easier and less expensive. Also, it may be integrated in a large-onset UWB system. The actual design can be improved using digitally controlled RF switches rather than 0 Ω resistors to select between *A1-A2-A3*. A higher permittivity substrate with better performance at higher frequency can be used to reduce the B-UWB-PG dimensions and jitter. The presenting solution has a lower jitter than the commercial HL9200 solutions, and from Figures 5.12 and 5.13, it may easily respect the UWB regulations.

This pulse generator solution can be used as an emitted stage of the two readers presented in Chapter 3 and 4. It will decrease the reader realization cost and the system jitter. The PRF of the pulse affects the resulting PSD/MHz, and the readers presented in Chapters 3 and 4 have a PRF of only 1 MHz. We note that this will result in a lower PSD/MHz for the UWB regulations, and therefore, a further amplification stage may be added at the output of the pulse generator.

In the following section, the designed pulse generator will first be used to read chipless tags, with the high-performance DSO in the reception stage and then with the reader presented in Chapter 4.

## 5.3. Measurement of a UWB pulse generator frequency-coded tag

The designed B-UWB-PG was tested to read the REP frequency-coded tag introduced in Vena, Perret and Tedjini [VEN 13a] and used along with the entire book to validate the different prototypes of the presented reader.

The B-UWB-PG was tested with two different configurations. The first exploits a single output port of the B-UWB-PG with the HMC788 gain block amplifier as the final stage. It is shown in Figure 5.14 and discussed in section 5.2.5.1. The second configuration is discussed in section 5.2.5.2. It makes use of the wideband combiner ZFRSC-123 from Mini-Circuits to combine the two single outputs of the B-UWB-PG to create a monocycle. The HMC788 was used as a final amplifier to compensate the combiner insertion loss.

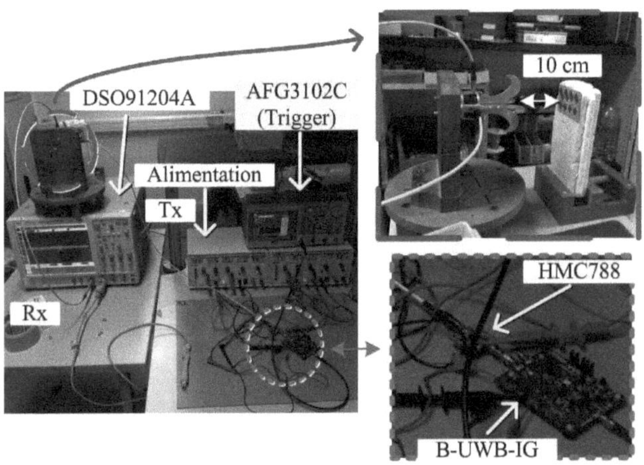


**Figure 5.14.** *Bench of tag measurements in a practical environment with the DSO DSO91204A in reception. The function generator AFG3102C is used as a trigger for the B-UWB-PG. One single output of the B-UWB-PG is used as the emission stage, along with the gain block amplifier HMC788. The B-UWB-PG is settled for an FWHM of about 90 ps, as shown in Figure 5.10. The tag is placed at a distance of 10 cm from the dual-access Satimo QH2000 antenna. The measurement is in cross-polarization, and the DSO averaging is 64. For a color version of this figure, see www.iste.co.uk/garbati/chipless.zip*

The B-UWB-PG was first tested with the high-performance DSO and then with the optimized reader version discussed in Chapter 4.

### 5.3.1. *Measurement results of the pulse generator*

The B-UWB-PG was used as the emission stage of an IR-UWB reader realized with test equipments. The DSO Agilent DSO91204A was in reception, while the antenna was the dual-access Satimo QH2000 with the tag placed at 10 cm. The bench is shown in Figure 5.14. The B-UWB-PG was configured with a single output amplified with the HMC788 and with an FWHM of about 90 ps, as shown in Figure 5.10. The DSO was settled with an averaging of 64, and the B-UWB-PG was triggered with the AFG AFG3102C. The measurement results are shown in Figure 5.15. All the measured tag frequency peaks, which correspond to the first mode resonant frequencies of the tags, are in correct positions.

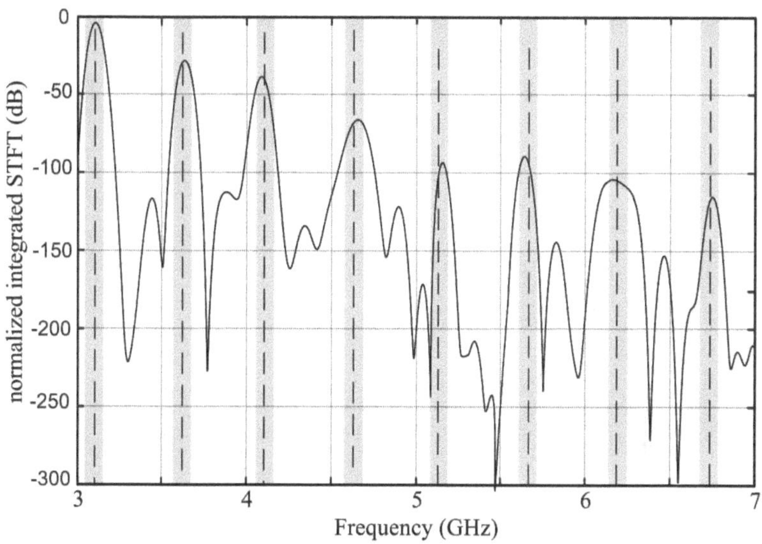


**Figure 5.15.** *Tag measurement result of the bench shown in Figure 5.14. All the tag frequency peaks are in the expected position (resonant frequencies)*

The same B-UWB-PG configuration was tested with the optimized reader version presented in Chapter 4, which was used in place of the DSO. Because the reader was designed around the HL9200 that has a higher output power, in the reader reception, the LNA Mini-Circuits TVA-82-213A+ with one attenuator of 18 dB was connected. The external LNA decreases the

ADC quantization noise (Chapter 4, section 4.2), resulting in higher reader ADC equivalent bit number. The LNA has a typical gain of 25 dB between 0.8 and 21 GHz; by adding the attenuator, the corresponding gain is reduced to 7 dB, but sufficient to retrieve the tag ID. To avoid the use of this external preamplifier, it is sufficient to increase the number of ADC bits (currently, they are 9 over 1.5 $V_{p\text{-}p}$) or to use the same driver employed in the reader version presented in Chapter 3 to add gain at the output of the S/H amplifier. The corresponding bench is shown in Figure 5.16, in which the reader was settled with the maximum averaging of 512. The corresponding measurement results are shown in Figure 5.17. All the tag frequency peaks are in the expected position (resonant frequencies). The results of the last peak are less accurate, which is due to the use of the external preamplifier and the pulse generator spectrum that shows attenuation at higher frequencies.

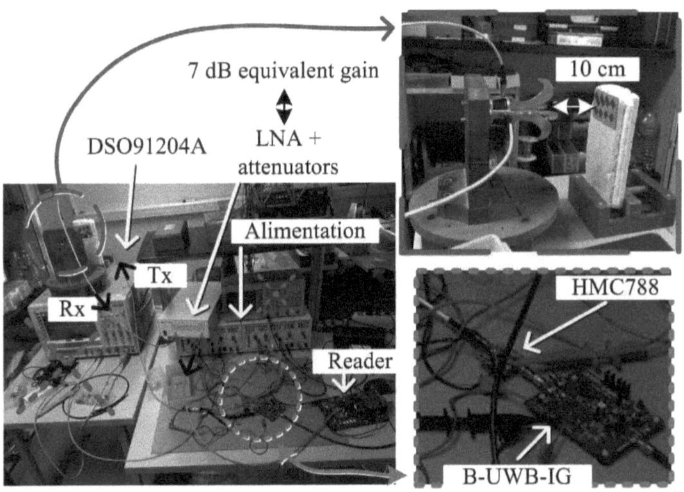


**Figure 5.16.** *Bench of tag measurements in a practical environment with the optimized reader proposed in Chapter 4 in reception. In reception, the LNA Mini-Circuits TVA-82-213A+ with one attenuator of 18 dB are interposed between the reader RF input and one port of the antenna. One single output of the B-UWB-PG is used as the emission stage, along with the gain block amplifier HMC788. The B-UWB-PG is settled for an FWHM of about 90 ps, as shown in Figure 5.10. The tag is placed at a distance of 10 cm from the dual-access Satimo QH2000 antenna. The measurement is in cross-polarization, and the reader averaging is 512. For a color version of this figure, see www.iste.co.uk/garbati/chipless.zip*

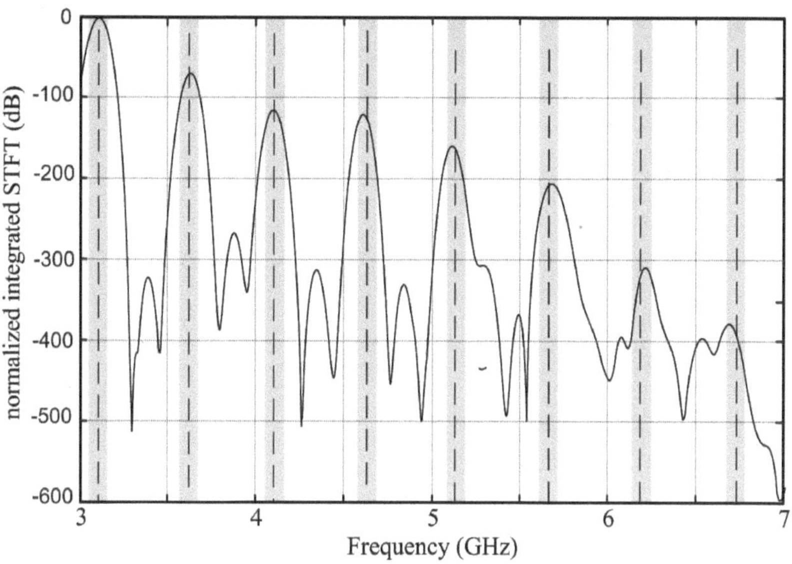


**Figure 5.17.** *Tag measurement result of the bench shown in Figure 5.16. All the tag frequency peaks are in the expected position (resonant frequencies). The high-frequency peak apexes are less accurate, due to the use of an external preamplifier and the pulse generator spectrum that shows attenuation at higher frequencies*

The second configuration for the B-UWB-PG was based on the monocycle creation shown in Figure 5.12. The same measurements with the DSO and the optimized reader presented in Chapter 4 were performed. With respect to the bench shown in Figures 5.14 and 5.16, the only difference was the B-UWB-PG configuration, and therefore, they are not shown. The measurement results are shown in Figure 5.18. Both DSO and reader approach show correct reading results.

## 5.3.2. *Conclusion*

This section has presented the potentiality of the designed B-UWB-PG as the emission stage of a UWB chipless RFID reader based on the IR-UWB approach. The measurements were performed in turn with a high-performance DSO and the optimized reader presented in Chapter 4 in reception in a practical environment. The two reading systems were able to read the tag introduced in Vena, Perret and Tedjini [VEN 13a], where the reader needed an external preamplifier of only 7 dB of gain to reduce the reader ADC quantization noise. The use of either an ADC with higher resolution (bit quantity) or a driver with a settled gain between the reader S/H amplifier and the analog ADC input will free up the reader by using the external preamplifier.

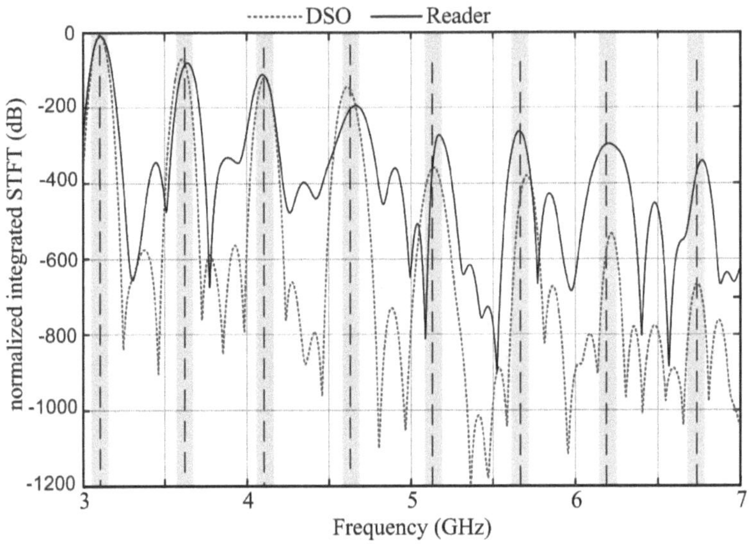


**Figure 5.18.** *Tag measurement result with a second B-UWB-PG configuration. All the tag frequency peaks are in the expected position (resonant frequencies). For a color version of this figure, see www.iste.co.uk/garbati/chipless.zip*

## 5.4. UWB antenna design

The readers presented in Chapters 3 and 4 use commercial ultra-large-band antennas, such as the Satimo QH2000. It is an Open-Boundary Quad-Ridge Horns antenna, operating in the bandwidth of 2–32 GHz, with a boresight gain between 10 and 14 dBi in the UWB of 3.1–10.6 GHz. The cost of this type of antenna is approximately 6 k€. For the deployment of the chipless technology, the reader should be less expensive. A printed antenna may represent a good candidate, where the feeding network should be simple to realize. An unbalanced antenna access point, such as an SMA connector, is preferable because it is compatible with the microstrip technology often used in a PCB layout.

Many different antennas can be used at the scope. First, a selection based on the polarization has to be done. Because the cross-polarization principle is used, the reader should be equipped with two linear polarized antennas placed in cross-configuration, aiming to decrease the coupling as much as possible. In a UWB chipless RFID system, the read range is limited, and because the low-power signal is reflected back from the tag and due to its linear behavior, the receiving antenna should have a high gain to maximize the link budget (see equation [2.7] in Chapter 2). The transmitting antenna may have a more isotropic character because both FCC and ETSI regulations [FCC 16, ETS 17] discriminated the maximum emitted power with the EIRP. On the contrary, a lower gain may generate a higher antenna coupling (leakage); hence, both the transmitting and receiving antennas should preferably have a high gain.

Among the planar printed UWB linear polarized antennas, the most commonly used are the monopole, dipole, Bow-Tie, Log-Periodic and Vivaldi. The monopole and dipole antennas have a low gain that is hardly higher than 5 dBi in all the band of interest (3.1–10.6 GHz) [RAM 16c]. The Bow-Tie antenna is composed of two triangular metal sheets, fed with a classical aperture coupling technique employing a broad-band stub. It has a quasi-omnidirectional radiation pattern along the H-plane and therefore a limited gain [WIE 09]. The Log-Periodic antenna has high bandwidth. It is a multi-resonant antenna and therefore has a variable group delay over its bandwidth. This makes reading time-based tags difficult because the antennas tend to spread out the transmitting or receiving UWB pulse, creating ringing effects [WIE 09]. The Vivaldi antenna is a frequency-independent antenna and therefore has reduced ringing effects. It has high gain and bandwidth and can be easily realized.

A Vivaldi antenna was fabricated on a high-performance laminate Rogers R03003 0.76 mm substrate, which has a relative dielectric constant of 3 and a dissipation factor tan $\delta$ of 0.0010 at 10 GHz. For the design, CST Studio Suite 2014 was used. The antenna has a balance radiative structure on the bottom formed of an exponential tapered slot line and an unbalanced fed structure based on microstrip. According to Figure 5.19, their coupling arrangement is composed of the radial stub on the top layer, coupled with the circular cavity on the bottom. Table 5.2 shows the corresponding design dimensions, and Figure 5.19 shows the CST design. The exponential tapered slot line has been modified during design to reduce the overall antenna dimension; therefore, its profile does not comply with the well-known equations proposed in Kahar, Mandal and Mandal [KAH 15].

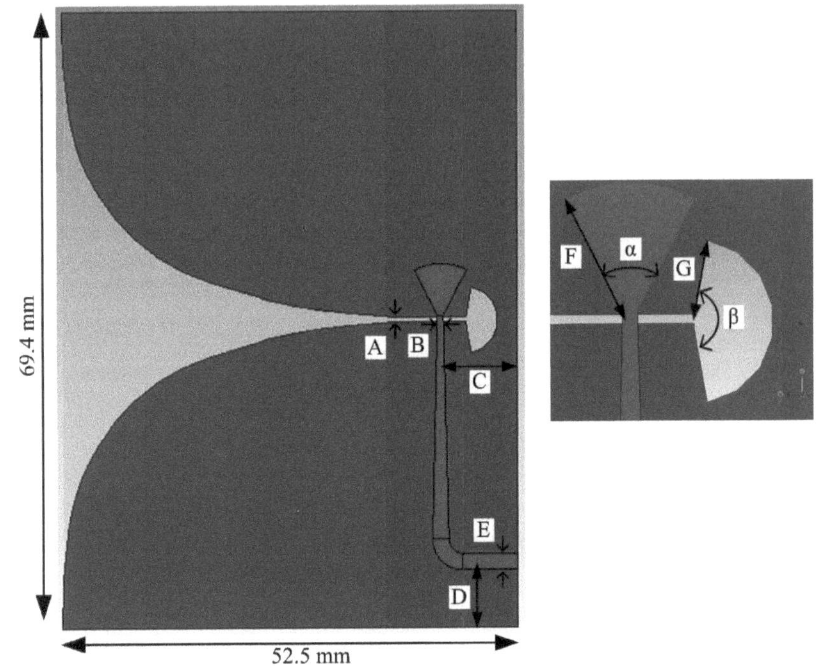


**Figure 5.19.** *Realized Vivaldi antenna*

| A | B | C | D | E | F | G | α | β |
|---|---|---|---|---|---|---|---|---|
| 0.42 mm | 0.70 mm | 8.92 mm | 7.48 mm | 1.89 mm | 6.52 mm | 3.67 mm | 0.304π rad | 0.885π rad |

**Table 5.2.** *Corresponding dimensions of Figure 5.19*

A photograph of the realized antenna and the measurement of its reflection coefficient are shown in Figures 5.20(a) and (b), respectively. The reflection coefficient is lower than –10 dB in the entire tested bandwidth 3.1–10 GHz. Figure 5.21 shows the simulated far-field gain at 5.5 GHz, with a maximum of 7.33 dBi. Figure 5.21(a) shows the polar chart in the E-plane. It shows an aperture of 49° and a side lobe level of –8.9 dB. Figure 5.21(b) shows the polar chart in the H-plane, with an aperture of 92° and a side lobe level of –7 dB.

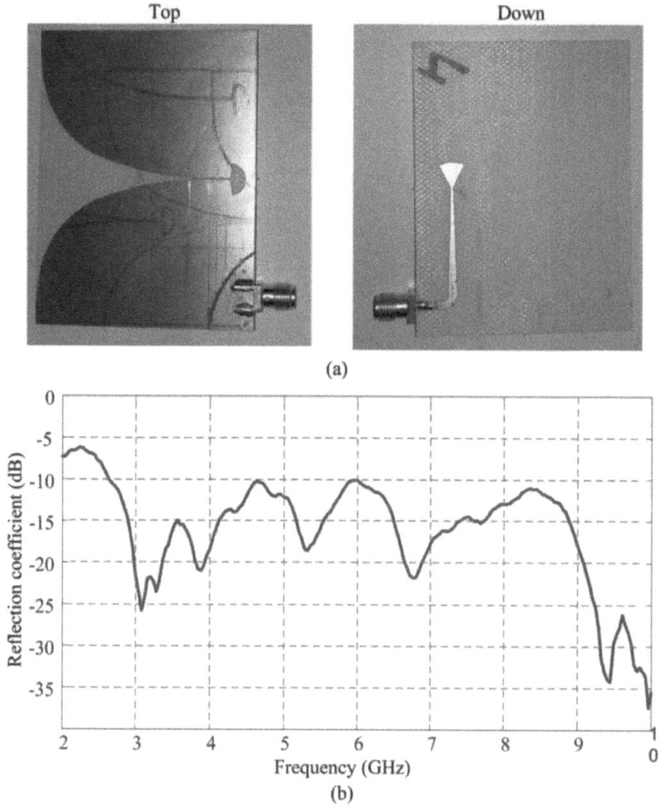


**Figure 5.20.** *Photograph and reflection coefficient measurement of the realized UWB Vivaldi antenna whose design is shown in Figure 5.19*

## 5.5. Conclusion

The design, realization and characterization of a low-jitter UWB pulse generator are described in section 5.5.2. It is based on COTS components and represents a low-cost solution. The retail cost of its main components after tax varies from 384 € to about 197 €. The more expensive solution provides a fine-tuning of the output pulse width, which is not possible for the latter that misses the delay generator chip. A prototype of the more expensive solution was realized on FR-4 and characterized. It shows a low random jitter of 2 ps standard deviation over 1,000 cycles. To comply with the UWB regulations, some shaping network configurations have been proposed. Its design has been facilitated exploiting the pulse generator flexibility. It may be used as the emitter stage of the UWB chipless RFID readers designed in Chapters 3 and 4, as demonstrated in section 5.3.

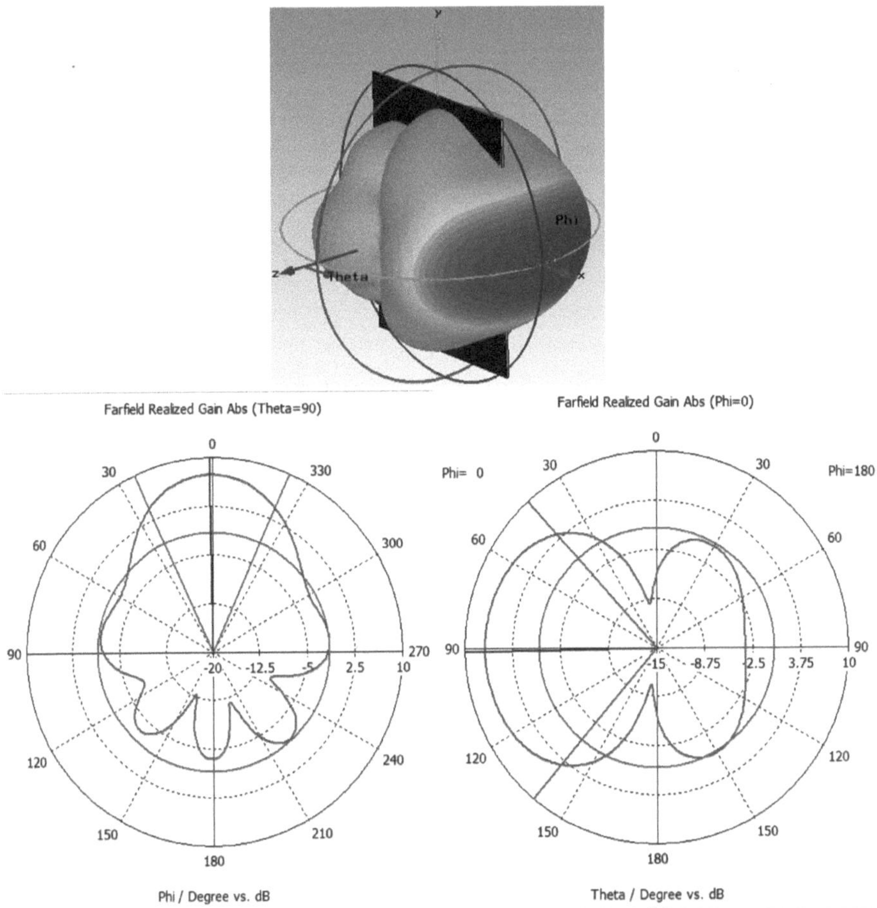


(a) (b)

**Figure 5.21.** *Simulated far-field gain at 5.5 GHz of the antenna shown in Figure 5.19. a) Polar chart of the gain in the E-plane, and b) polar chart of the gain in the H-plane. For a color version of this figure, see www.iste.co.uk/garbati/chipless.zip*

A planar Vivaldi UWB antenna has been realized and described in section 5.4. It shows a reflection coefficient lower than –10 dB in the entire bandwidth 3–10 GHz. It was designed to have an unbalanced alimentation structure so as to be used with the proposed readers.

In the following chapter, the problem of reading UWB chipless RFID tags, independently from their orientation, will be addressed. A technique based on polarization diversity will be proposed. A reading system based on test equipment was realized as a proof of concept, and then the same principle was applied to the reader presented in Chapter 4.

# 6

# UWB Chipless RFID Reading System Independent of Tag Orientation

## 6.1. Introduction

UWB chipless RFID tags must be aligned with reader antennas to maximize the reading capability when a cross-polarization approach is used. The reading therefore depends on tag orientation. In this chapter, we propose a reading technique based on polarization diversity from the reader part, in order to obtain a reading that is rather independent of tag orientation that maintains a cross-polarization configuration. In the first part of the chapter, some measurements of chipless RFID tags with state-of-the-art test equipment are shown. Then, to read tags independently of their orientation, a modified version of the optimized reader presented in Chapter 4 is detailed. This chapter is organized as follows:

– Section 6.2 introduces the problems of reading tags for different orientations from a theoretical perspective.

– Section 6.3 presents the potentiality of a four-port state-of-the-art VNA to perform balanced measurements. The proposing reading technique is introduced and balanced measurements are made with the VNA to read the tags for different orientations in a practical environment.

– Section 6.4 demonstrates the feasibility of the proposed technique, with a reduced number of measurements (only two), to be used in practical applications.

– Section 6.5 presents the application of the proposed method to the optimized reader presented in Chapter 4, with the addition of some external hardware components. Measurements of tags in a practical environment are also demonstrated.

– Section 6.6 draws the conclusion of this chapter.

## 6.2. Principle of operation

One of the major drawbacks of UWB chipless RFID systems is polarization dependence. Most of the tags proposed in the literature have to be specifically oriented with the reader, working in linear polarization mode and using high directive antennas to maximize the read range [PER 14]. The tags as well as the reader are anisotropic and have some discrete directions of maximum radiation. Thus, tags must be correctly oriented with the reader to maximize the reading. To attain a UWB chipless RFID system independent of mutual orientation, dedicated work may involve either the reader or the tag design. Chipless UWB tags with circularly polarized antennas [SHE 12] or tags independent of polarization [VEN 12b] have been proposed; however, these tags have generally reduced performance when compared to linear cross-polarized tag [VEN 13a].

A typical scenario of a tag-reading process is shown in Figure 6.1. The tag is read in a cross-polarization configuration; therefore, the E-planes of the reader transmitting and receiving antennas are orthogonal, as shown in the figure. The tag is supposed to be aligned with the reader antennas in the direction of maximum backscattering RCS. Then, the tag can be interrogated with different reader antenna orientations, defined by the angle $\alpha_T$ (defined in Figure 6.1), to characterize the response of the tag. $\alpha_T$ is the angle of the intersection between the E-plane of the reader transmitting antenna and the Cartesian coordinate system (x,y) defined in the tag plane. A variation of $\alpha_t$ corresponds to a rotation of the reader antennas. The tag position is defined with β, which is the angle between the x axis and the tag direction. This direction is defined by the straight line parallel to its longer size and crossing the origin of the Cartesian coordinate system (x,y). The tag shown in Figure 6.1 is the cross-polarized tag presented in [VEN 13a], which is used to characterize the different reader prototypes described in this chapter.

A model of the tag was realized in CST Studio Suite and simulated by varying $\alpha_T$ between 0° and 90°. For simplicity, the tag was placed at β = 0°, as shown in Figure 6.2. The tag response is shown in Figure 6.3, and the positions of the eight tag resonant frequencies versus the angle are shown in gray dashed lines. The STFT algorithm was employed, in post-processing with Matlab®, to remove the tag structural mode contribution from the simulated tag response [RAM 16b]. The tag shows a symmetric response at 45°: the tag maximum response is for $\alpha_t$ = 0° and 90°, and the minimum for $\alpha_t$ = 45°. The tag ID was correctly retrieved with $\alpha_t$ between 0° and 35°, with results similar to $\alpha_t$ between 55° and 90°. For other directions at 45°, the reading was difficult because the tag response amplitude was reduced by several decibels with respect to more favorable cases (0° and 90°).

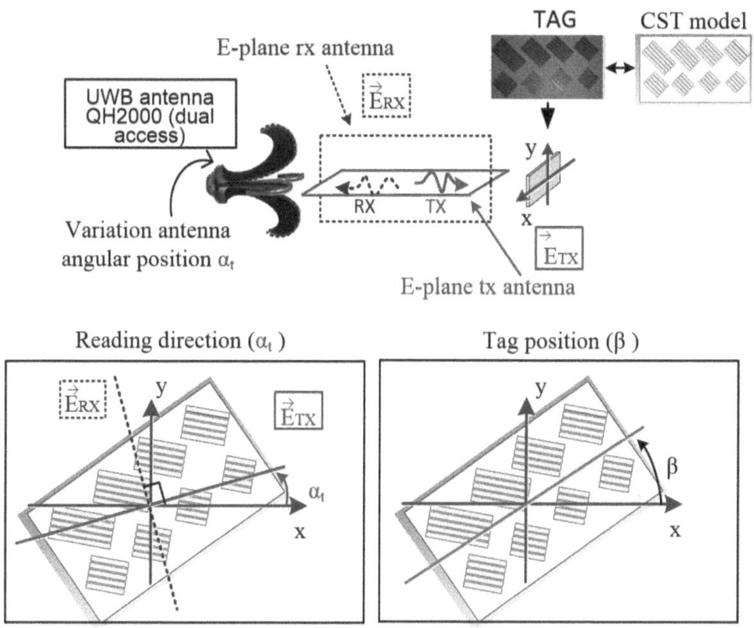


**Figure 6.1.** *Scenario of a tag-reading process in cross-polarization configuration, varying the direction of interrogation ($\alpha_T$). $\alpha_T$ is the angle of the intersection between the E-plane of the reader transmitting antenna and the Cartesian coordinate system (x,y) defined in the tag plane. A variation of $\alpha_t$ corresponds to a rotation of the reader antennas. The tag position is defined with $\beta$, which is the angle between the x axis and the tag direction. This direction is defined by the straight green line parallel to its longer size and crossing the origin of the Cartesian coordinate system (x,y). For a color version of this figure, see www.iste.co.uk/garbati/chipless.zip*

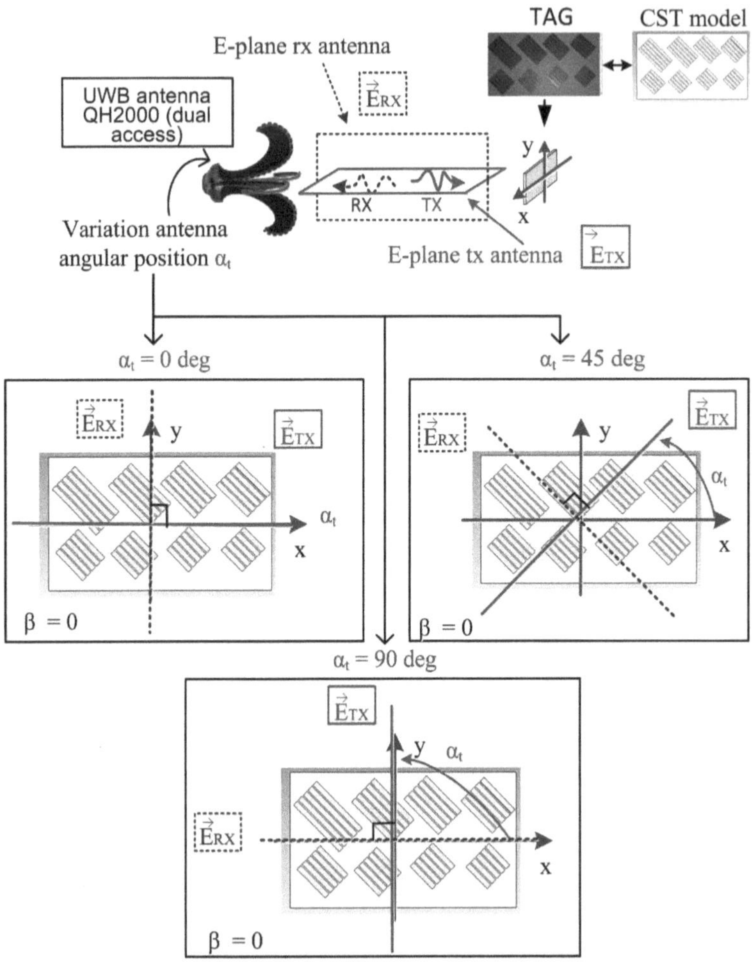


**Figure 6.2.** *Scenario of reading a tag in a cross-polarization configuration by varying the direction of interrogation ($\alpha_t$) with the tag position $\beta$ = 0°. A CST Studio Suite model of the tag was realized and simulated by varying $\alpha_t$ between 0° and 90°. The tag was the cross-polarized tag presented in [VEN 13a]. It is characterized by eight resonant frequencies in the bandwidth of 3–7 GHz. For a color version of this figure, see www.iste.co.uk/garbati/chipless.zip*

The simulation performed covers an $\alpha_t$ variation in the first and third quadrants of the x,y Cartesian coordinate system (the angle range of 0–90° being equivalent to the range of 180–270°). The behavior of the tag for the second and fourth quadrants, with $\alpha_t$ between 90° and 180° (equivalent to 270–360°) gives almost the same results due to the tag shape. Thus, the response of a cross-polarized tag is periodic at 90°. This has been reported in the literature and will be formalized in section 6.4.

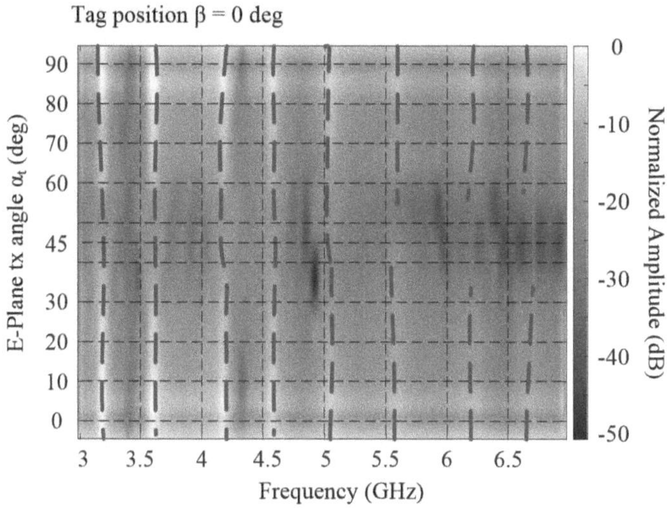


**Figure 6.3.** *Simulation result of the scenario shown in Figure 6.1, with the tag placed at $\beta$ = 0°. The STFT algorithm was employed in post-processing with Matlab® to remove the tag structural mode contribution from the simulated tag response. The tag has a symmetrical behavior around the minimum response of $\alpha_t$ = 45°. The maximum is at 0 and 90°. The retrieved position of the eight resonant frequencies is represented by gray dashed lines. For a color version of this figure, see www.iste.co.uk/garbati/chipless.zip*

As has been shown, typically, cross-polarized tags have two preferential reading directions [VEN 12b] that maximize the antenna (tag-) mode cross-polarized RCS and allow the recovery of the tag ID under the best condition. They are orthogonal and therefore 90° distant and, as shown in Figures 6.2 and 6.3, they correspond to the case of $\alpha_t = \beta \pm n\, 90°$, where $n$ is an integer. There are two perpendicular directions of minimum tag response between them, as shown in Figure 6.4. When $\beta = 0°$, they are located at $\alpha_t = 45°$ and 135° ($\beta \pm n\, 45°$, where $n$ is an odd integer). The aim of this book is to improve the state of the art of UWB chipless RFID reader technology; therefore, the orientation problem has been studied from that perspective. The goal is to show that a reader can electrically rotate its interrogating signal, varying $\alpha_t$, while always receiving in the perpendicular direction (i.e. the reading is always in cross-polarization).

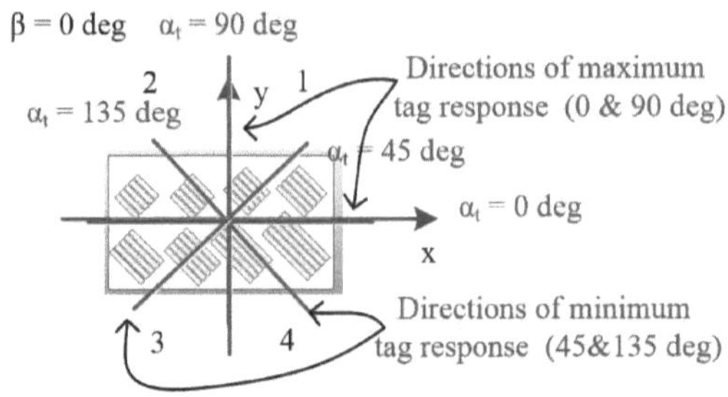


**Figure 6.4.** *Directions of the tag's maximum and minimum response for cross-polarization reading. For a color version of this figure, see www.iste.co.uk/garbati/chipless.zip*

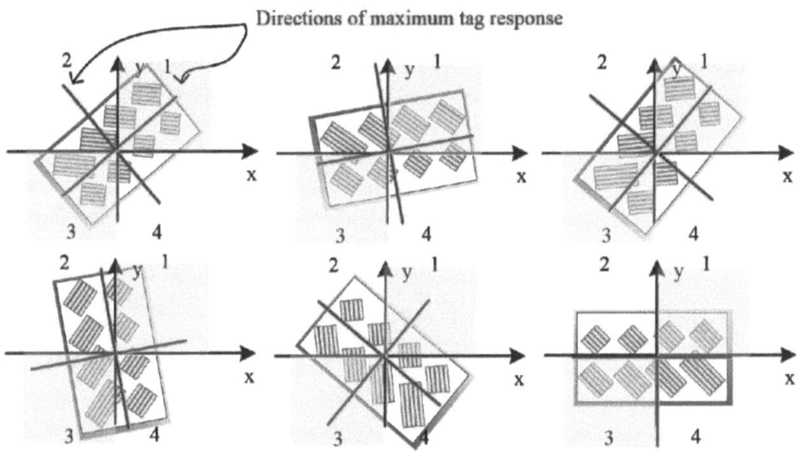


**Figure 6.5.** *Reading process with $\alpha_t$ varying in the interval [0°–90°] corresponding to a sweep in the first and third quadrants of the Cartesian coordinate system (x,y). Regardless of the position of the tag, at least for one time, the tag is interrogated along with one of its two preferential reading directions. For a color version of this figure, see www.iste.co.uk/garbati/chipless.zip*

To read the tag independently from its orientation, one can sweep in the interval of either [0–90]° or [90–180]°, due to the tag response symmetry, as shown in Figure 6.5. This ensures the reading of the tag under the best condition for at least one value of $\alpha_t$, maximizing the antenna (tag-) mode cross-polarized RCS. For simplicity, this chapter focuses on the interval [0–90]°. This approach has been demonstrated using the four-port Agilent VNA N5222A, in a balanced measurement configuration. The following section introduces the principle of operation for VNA balanced measurements.

## 6.3. VNA balanced measurements

### 6.3.1. *Variation of the transmitting signal direction $\alpha_t$*

The N5222A VNA is equipped with four physical ports, which allows balanced measurements to be performed in true mode (real balanced signals) [KEY 16b]. The four ports of the VNA are grouped into two logic ports 1 and 2, where the first logic port is composed of physical ports 1 and 3, while the second logic port is composed of physical ports 2 and 4, as shown in Figure 6.6. The signals at the two logic ports have a common and differential component, which may be defined, respectively, as in equations [6.1] and [6.2] for logic port 1 and as in equations [6.3] and [6.4] for logic port 2.

$$V_{LP1-Common} = \frac{V_1+V_3}{2} \quad [6.1]$$

$$V_{LP1-Diff} = \frac{V_1-V_3}{2} \quad [6.2]$$

$$V_{LP2-Common} = \frac{V_2+V_4}{2} \quad [6.3]$$

$$V_{LP2-Diff} = \frac{V_2-V_4}{2} \quad [6.4]$$

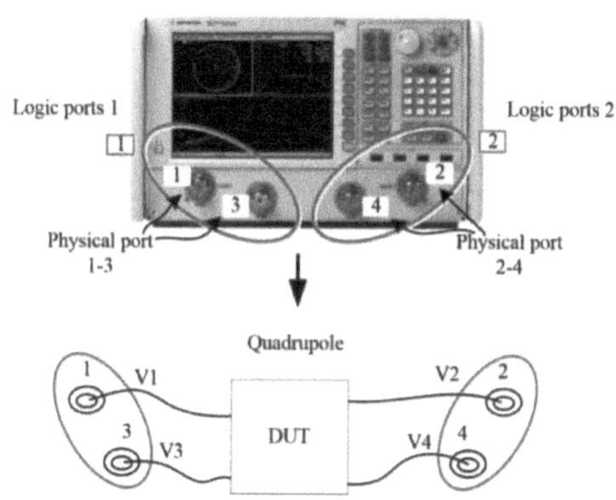


**Figure 6.6.** *Organization of the VNA four physical ports in a balanced measurement configuration. They are arranged with two logic ports, each associated with two physical ports. Logic port 1 is associated with physical ports 1 and 3, whereas logic port 2 is associated with physical ports 2 and 4*

It is possible to perform common-to-common (c–c), common-to-differential (d–c), differential-to-differential (d–d) and differential-to-common (c–d) S-parameter measurements, from logic port 2 to logic port 1, and vice versa, where c and d stand for common and differential, respectively. A balanced signal, which has a null differential component, is composed of two signals with equal amplitude and phase (equations [6.2] and [6.4]). If the common component is null, then the two signals are equal in amplitude and opposite in phase (equations [6.1] and [6.3]). In addition to balanced-to-balanced measurements, it is possible to perform single-to-balanced and balanced-to-single measurements. From Figure 6.6, the only difference is that one of the two logic ports is composed of only one physical port. Compared to balanced-to-balanced measurements, other measurements have some limitations in true mode, as has been briefly explained.

Satimo QH2000 is a dual-access antenna, which can therefore be connected to one of the two logic ports once in a balanced configuration. The antenna connected to logic port 2 (physical ports 2 and 4) is shown in Figure 6.7. Each antenna component, vertical and horizontal, will transmit a linearly polarized EM signal along with the respective E-planes. The resulting signal toward the tag will be the vector sum between the two linearly polarized signals, as shown in Figure 6.7. To avoid an interrogating signal with elliptical polarization, the two exiting components of the antenna, $V_4$ and $V_2$, have to be in phase or out of phase. If the exiting signals of logic port 2 are common signals, then the two components have the same phase. Therefore, the interrogating signal will be a linearly polarized one in the first and third quadrants of the Cartesian coordinate system (x,y) (see Figure 6.7). The transmitting angle $\alpha_t$ will be equal to 45° because the two components also have the same amplitude. If the transmitting signal is differential, their two components will be out of phase and $\alpha_t$ will be equal to 135°. As described at the end of the previous section, the selected interval of variation of $\alpha_t$ to interrogate the tag was [0–90]°. Therefore, to vary $\alpha_t$ in that interval, keeping a linearly polarized transmitting signal, the amplitudes of the two components, $V_4$ and $V_2$, must be varied, while keeping the same phase.

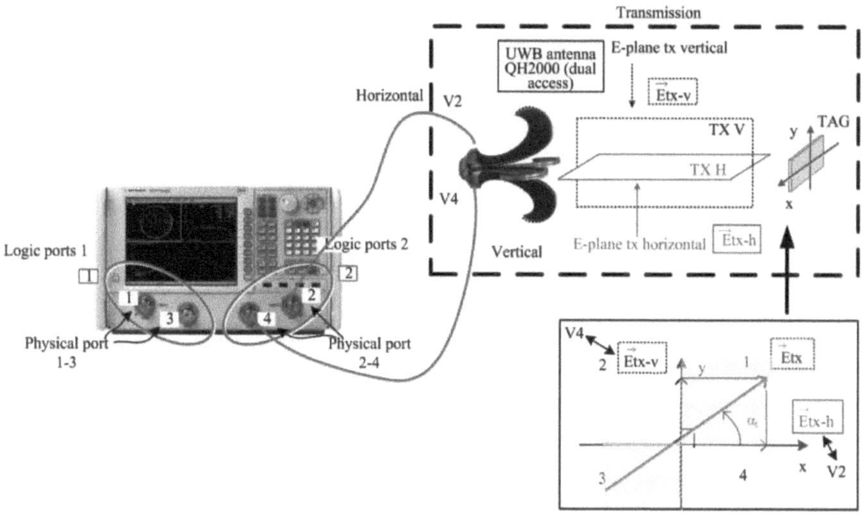


**Figure 6.7.** *Connection between the VNA and a dual-access antenna. Using logic port 2 of the VNA in transmission with a balanced configuration, the transmitting signal toward the tag will be the vector sum of the two linearly polarized signals in the two orthogonal E-planes of the antenna. For a color version of this figure, see www.iste.co.uk/garbati/chipless.zip*

The VNA allows the setting of a power offset between the two signals $V_4$ and $V_2$ of a balanced port, maintaining the same phase relationship. The user can select the amount of power offset, whereas the sum of the transmitted power of the two single ports will be constant. Thus, it is possible to vary $\alpha_t$ in transmission in the interval [0°–90°]. In a more general application, this option is used to compensate for the mismatch between cables and the PCB layout when measuring a chip characteristic as the common-mode rejection ratio (CMRR) of an operational amplifier. The power offset between the two signals, $V_4$ and $V_2$, is

$$P_{offset} = 20log10(\frac{V_4}{V_2}). \quad [6.5]$$

The angle $\alpha_t$ depends on the amplitude of the two electric fields along the two E-planes of the antenna, $\vec{E}_{tx-v}$ and $\vec{E}_{tx-h}$. These electric-field amplitudes are directly proportional to the amplitudes of $V_2$ and $V_4$. Thus, $\alpha_t$ can be expressed as

$$\alpha_t = \arctan(\frac{V_4}{V_2}). \quad [6.6]$$

Combining equations [6.5] and [6.6], the relationship between the power offset of the two signals, $V_4$ and $V_2$, and the transmitting signal direction $\alpha_t$ is

$$\alpha_t = \arctan\left(10^{\frac{P_{offset}}{20}}\right). \qquad [6.7]$$

Table 6.1 presents some discrete values obtained using the aforementioned equation. As can be observed, a power offset between ±15 dB is sufficient to cover almost the entire range [0°–90°].

| **Relationship between $\alpha_T$ and $P_{offset}$** | | | | | | | | | |
|---|---|---|---|---|---|---|---|---|---|
| $\alpha_t$ (°) | 90 | 80 | 70 | 60 | 45 | 30 | 20 | 10 | 0 |
| $P_{offset}$(dB) | ∞ | 15 | 8.8 | 4.7 | 0 | –4.7 | –8.8 | –15 | –∞ |

**Table 6.1.** *Relationship between $\alpha_T$ and $P_{offset}$ using equation [6.7]*

It is possible to make common-to-single (s–c) S-parameter measurements from logic port 2 to logic port 1, i.e. Ssc12. The transmitting balanced logic port 2 is composed of two physical ports 2 and 4. The receiving logic port 1 is single and composed of only physical port 1. Thus, the transmitted balanced signal (logic port 2) has a differential component equal to 0 and presents only a common component. This means that the two exiting signals of the transmitting antenna, $V_2$ and $V_4$, are in phase and equal in amplitude. Their power offset is 0 dB, and according to Table 6.1, the transmitted linearly polarized signal has an angle $\alpha_t$ of 45°. Then, it is possible to vary the power offset between the two components, according to Table 6.1, to modify $\alpha_t$.

The true mode must be selected, otherwise the VNA will emulate only a balanced measurement. The VNA will only perform single-to-single measurements on the three physical ports 2, 4 and 1. Finally, the VNA will arrange the results to retrieve the results as for a real balanced measurement. Therefore, the true-mode option is required for these experiments.

Another VNA option that must be selected is called the *offset as fixture*. It will not take into account, or more exactly compensate for, the imposed power offset on the balanced logic ports in post-processing to retrieve the measurement result. According to Table 6.1, varying the power offset of the transmitting signal of logic port 2 will create two signals, $V_2$ and $V_4$, with the same phase but different amplitudes. The resulting transmitting signal toward the tag will be still linearly polarized, but from equations [6.3] and [6.4], a differential component will be created besides the common one. The VNA is settled with common-to-single analysis; therefore, it will only make calculation on the common component of the transmitting signal, expressed using equation [6.3], to yield the measurement result. When the option *offset as fixture* is selected, the VNA will not take into account the imposed power offset variation from the calculation, and the transmitting signal is considered purely common in the calculation. Indeed, independently from the selected power offset, the transmitting balanced signal has the same power.

The variation of the transmitting signal between $\alpha_t$ = 10° and 80°, which corresponds to a power offset variation $P_{offset}$ from –15 to 15 dB (Table 6.1), was validated in an anechoic environment. The bench of the validation measurement process is shown in Figure 6.8. One Satimo QH2000 antenna in transmission is connected to the balanced logic port 2 of the VNA, corresponding to physical ports 2 and 4, and the other is connected to the VNA logical port 1 (physical port 1 only). This second antenna can be considered as a linearly polarized probe in reception, where only one of the two "Vivaldi" antennas forming the QH2000 is connected to the VNA, and the other connected to a 50 Ω SMA load. The rotation of the probe antenna (receiving antenna) was executed with an electrical engine, and the two antennas were separated by a distance of approximately 25 cm to recreate a typical UWB chipless RFID reading range.

The measurement results of the bench presented in Figure 6.8 are shown in Figure 6.9. In Figure 6.9(a), the power offset is 15 dB, corresponding to $\alpha_t$ = 80°. In Figure 6.9(b), the power offset is 0 dB, corresponding to $\alpha_t$ = 45°, whereas in Figure 6.9(c), the power offset is –15 dB, corresponding to $\alpha_t$ = 10°. These theoretical values are validated with the corresponding measurements shown in Figure 6.9.

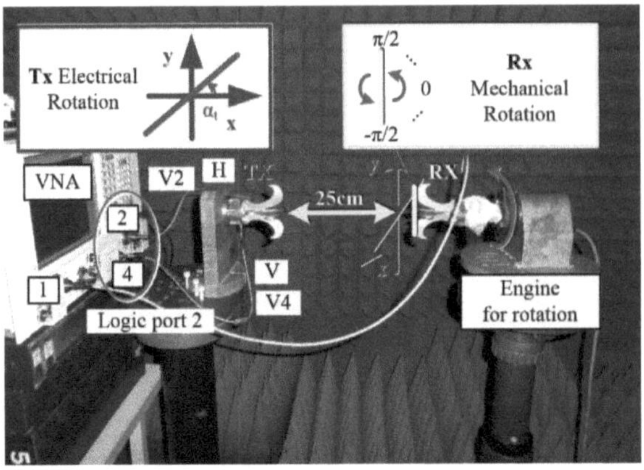


**Figure 6.8.** *Validation of the transmission Ssc12 balanced measurement in an anechoic environment. A power offset $P_{offset}$ is used in transmission (logic port 2) to turn the Tx linearly polarized signal with the VNA option power offset as fixture. One Vivaldi component of the Satimo QH2000 antenna is used as a probe in reception (single port 1), which is mechanically rotated to analyze the characteristics of the transmitted signal. For a color version of this figure, see www.iste.co.uk/garbati/chipless.zip*

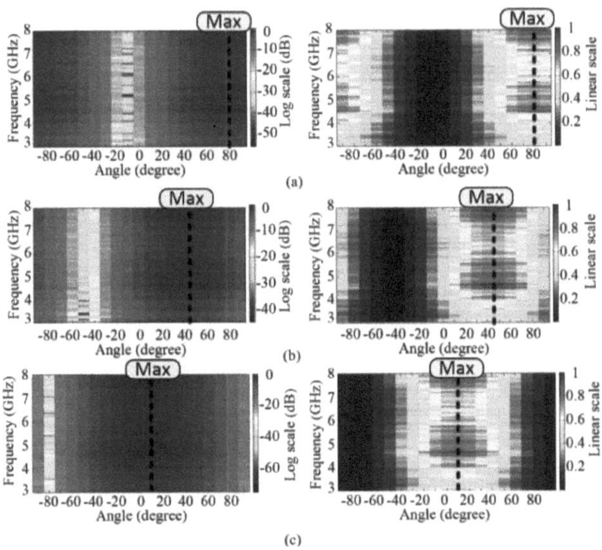


**Figure 6.9.** *Validation of the transmission Ssc12 balanced measurement for power offsets of a) 15 dB, corresponding to 80°, b) 0 dB, corresponding to 45° and c) −15 dB, corresponding to 10°. The scale is semi-logarithm on the left-hand-side 2D plots and linear on the right-hand side plots. For a color version of this figure, see www.iste.co.uk/garbati/chipless.zip*

### 6.3.2. *Isolating the cross-polarized component in reception*

In reception, the cross-component (related to the polarization of the transmitting signal) of the receiving signal from the tag must be measured. It corresponds at the EM field component with an angle equal to $\alpha_t$ + 90°. The receiving antenna may be another dual-access device connected to the physical ports 1 and 3. Then, it is possible to extract the tag contribution at $\alpha_t$ + 90° (cross-polarization) from the signal measured at the two ports of the antenna. To realize this, the clockwise system of reference rotation matrix can be applied at these two components to obtain the response in the desired direction.

The measurement setup of a reading system using a four-port VNA Agilent N5222A and two dual-access Satimo QH2000 wideband antennas is shown in Figure 6.10(a). One antenna is used in transmission and the other in reception. Ports 2 and 4 of VNA are connected to the transmitting antenna (Tx antenna), while ports 1 and 3 are connected to the receiving antenna (Rx antenna). As shown in the figure, the transmitting and receiving antennas have the same orientation; hence, the rotation matrix must be applied to an angle equal to $\alpha_t$, which is

$$\begin{pmatrix} V_{cross} \\ V_{co} \end{pmatrix} = \begin{pmatrix} \cos(\alpha_t) & -\sin(\alpha_t) \\ \sin(\alpha_t) & \cos(\alpha_t) \end{pmatrix} \begin{pmatrix} V_3 \\ V_1 \end{pmatrix}. \qquad [6.8]$$

where $V_1$ and $V_3$ are, respectively, the receiving signals on the VNA physical ports 1 and 3, which are connected to the vertical and horizontal components of the receiving antenna; $V_{cross}$ is the part of the receiving signal along the desired direction at $\alpha_t$ + 90° and $V_{co}$ represents the response along the direction defined by $\alpha_t$. From equation [6.8], $V_{cross}$ is

$$\text{Vcross} = cos(\alpha_t)V_3 - \sin(\alpha_t)\, V_1. \qquad [6.9]$$

### 6.3.3. *Measurements of cross-polarized tags*

A simple way to measure the two components of the receiving antenna, as shown in Figure 6.10(a), would be to set up the VNA with two common-to-single measurements Ssc12. The single logic port 1 is the physical port 1 for one measurement and the physical port 3 for the other. Unfortunately, this is not possible in a true-mode approach. Therefore, another option provides the execution of one common-to-common (Scc12) measurement and one common-to-differential (Sdc12) measurement. In reception, logic port 1 is composed of the two physical ports 1 and 3 and is treated as a balanced port. The measurements in reception of the common and differential components of the balanced receiving signal correspond to the two signals of equations [6.1] and [6.2]. From Figure 6.10(b), it is equivalent to a rotation of 45° of the receiving Cartesian coordinate system (x,y). Finally, the rotation matrix of equation [6.8] must be modified as in equation [6.10], to take into account the rotation of 45° (see Figure 6.10(b)) already performed, and applied to the measured signals $V_{LP1-Common}$ and $V_{LP1-Diff}$:

$$\begin{pmatrix} V_{cross} \\ V_{co} \end{pmatrix} = \begin{pmatrix} \cos(\alpha_t - 45) & -\sin(\alpha_t - 45) \\ \sin(\alpha_t - 45) & \cos(\alpha_t - 45) \end{pmatrix} \begin{pmatrix} V_{LP1-Common} \\ V_{LP1-Diff} \end{pmatrix}. \quad [6.10]$$

Then, $V_{cross}$ becomes

$$\text{Vcross} = cos(\alpha_t - 45) V_{LP1-Common} - \sin(\alpha_t - 45)\, V_{LP1-Diff}. \quad [6.11]$$

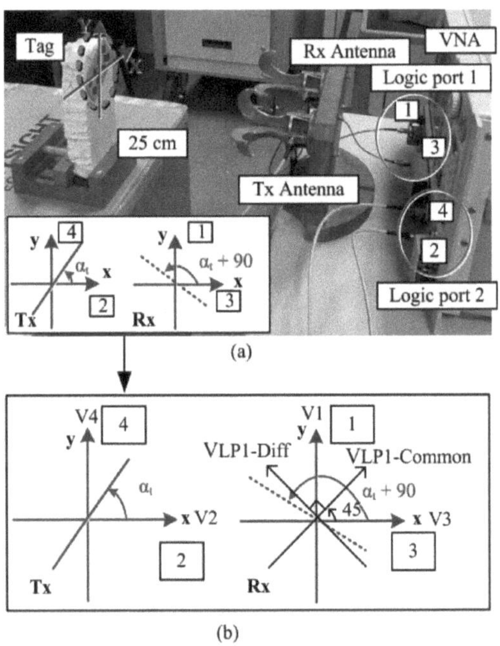


**Figure 6.10.** *a) Measurement setup of a reading system using a four-port VNA Agilent N5222A and two dual-access Satimo QH2000 wideband antennas. One antenna is used for transmission and the other for reception. b) Antenna configuration, where the common and differential components ($V_{LP1-Common}$, $V_{LP1-Diff}$) of the receiving signal are measured. For a color version of this figure, see www.iste.co.uk/garbati/chipless.zip*

This approach was validated in a practical environment, where the bench is shown in Figure 6.10(a). The tag was placed at different orientations (β) and, for each orientation, the transmitting signal direction was modified to retrieve the tag position ($\alpha_t$ = β). The power offset ($P_{offset}$) in transmission was selected according to Table 6.1. The measurement results are shown in Figure 6.11. The STFT approach was used, and the results are normalized. The tag orientation β was varied with the discrete positions: 0°, 30°, 45° and 90°. The tag was correctly decoded for all of its positions.

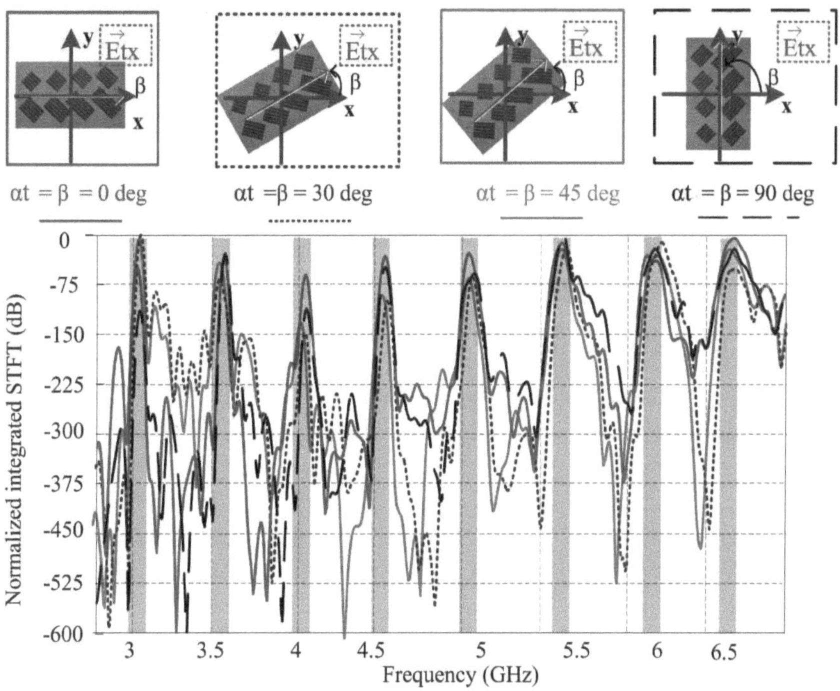


**Figure 6.11.** *Measurement of the cross-polarized tag presented in [VEN 13a], for different orientations (β), with respect to the bench presented in Figure 6.10. The VNA was settled with two measurements, Scc12 and Sdc12. The angle $\alpha_t$ varies in transmission to retrieve the tag position β, and equation [6.11] is used to retrieve the cross-polarized tag response. For a color version of this figure, see www.iste.co.uk/garbati/chipless.zip*

The proposed approach expects the variation of $\alpha_T$ in the interval [0°–90°], which is controlled by the power offset ($P_{offset}$) between the two physical ports 2 and 4 of the transmitting logic port 2. As will be demonstrated in the next section, in a practical approach, the backscattered signal from the tag must be evaluated only for two orientations $\alpha_T$ (one at an angle of 45° from the other) to retrieve the tag ID independently from its orientation. A GUI software in Matlab®, presented in Appendix 5, has been used to perform balanced measurements.

## 6.4. Tag measurements using the VNA: a simplified approach

### 6.4.1. *Practical approach*

The variation of antenna (tag-) mode cross-polarized RCS of a tag versus the angle of observation (transmitting signal direction) was characterized in [RAN 17] and simulated in the first section of this chapter, and the results are shown in Figure 6.3. The tag-mode cross-polarized RCS is a function of the difference between β and $\alpha_t$, which can be expressed using equation [6.12]:

$$\frac{\sigma_{\alpha_t-\beta}}{\sigma_{CROSS}} = |(1 - 2sin^2(\beta - \alpha_t))|^2 \qquad [6.12]$$

where $\sigma_{\alpha_t-\beta}$ is the actual RCS under that condition of measurement and $\sigma_{CROSS}$ is the RCS in the best case, i.e. when $\alpha_t$ = β ±*n* 90°, with *n* being an integer. The domain of the function is defined by $\beta - \alpha_T$. The transmitting angle $\alpha_t$ was chosen between 0° and 90°. The function is periodic at 90° and plotted in linear and semi-logarithmic scales, as shown in Figures 6.12 (a) and (b), respectively. The variation of the function amplitude around their maximum is 6 dB in the range of ±30°. Outside this range, $\sigma_{\alpha_t-\beta}$ decreases quickly with a theoretical dynamic variation of ∞ dB when the variation is of ±45°.

From equation [6.12], represented in a semi-logarithmic scale, for any two points in the function domain with a distance of 45°, at least one of the two will be inside the favorable range of ±22.5° around one maximum of the function. It corresponds to a maximum RCS variation of 3 dB with respect to the maximum $\sigma_{CROSS}$, which allows the tag ID to be retrieved without ambiguity. Because the tag position β is fixed, selecting two points at an angle of 45° between them means varying the transmitting interrogating angle $\alpha_t$ of 45°. Thus, the proposed approach presented in the previous section may be used. In such a case, only two variations of $\alpha_t$ will be imposed to retrieve the tag response (ID) regardless of its orientation. The sole constraints are: (1) the two $\alpha_t$ have to be 45° distant from each other and (2) from the two measurement results, a simple criterion must be applied to find which of the two contains the tag information (ID). For instance, Figure 6.13 shows the measurements of one tag using the two angles $\alpha_t = 0°$ and 45°. The tag is placed at the discrete positions $\beta = 0°$ and 45°. Figure 6.13 (a) shows the measurement results for the transmitting angle $\alpha_t = 0°$. When the tag is in the horizontal position ($\beta = 0°$), the measurement is completely successful, while when $\beta = 45°$, the results are inconsistent with the presence of significant secondary lobs around some of the tag resonances. This is because $\beta - \alpha_t = 45°$ that represents one minimum of equation [6.12] (see Figure 6.12). Instead, Figure 6.13(b) shows the measurement results varying $\alpha_t$ to 45°, and the previous case of failure becomes a completely successful reading.

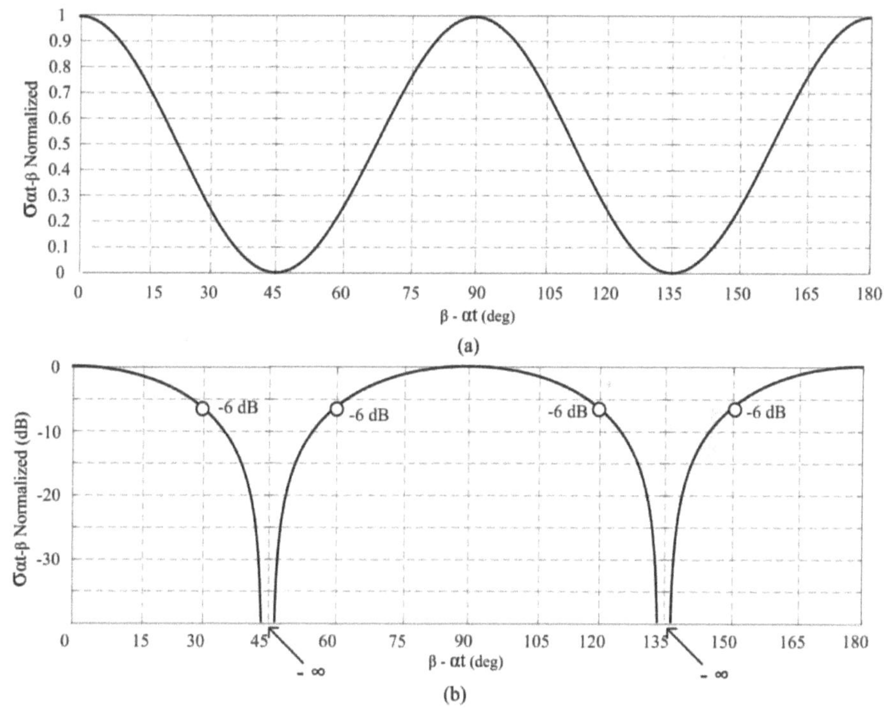


**Figure 6.12.** *Representation of equation [6.12] in a) linear and b) semi-logarithm scales. Variation of $\sigma_{\alpha_t-\beta}$ in the range of ±30°, around one maximum of the function, ensures a cross-polarized RCS dynamic variation of 6 dB*

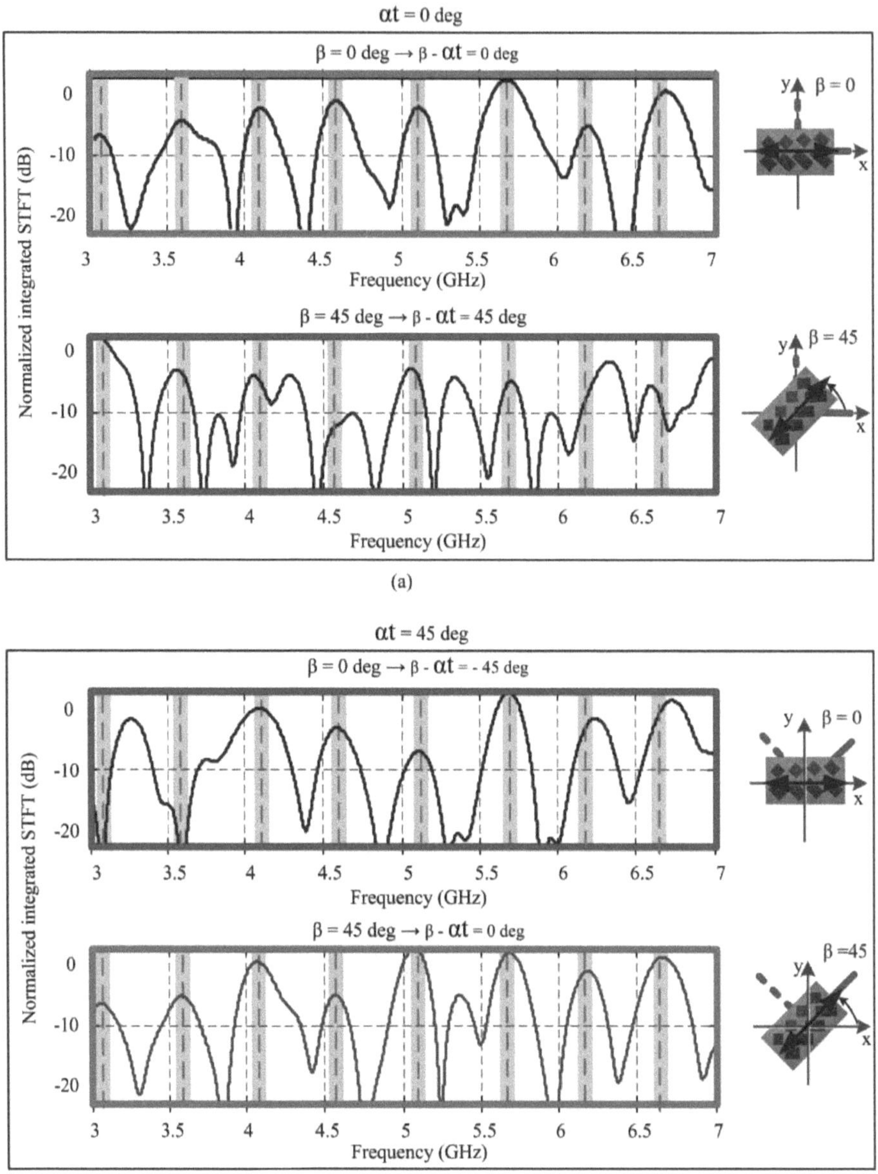


(b)

**Figure 6.13.** *Reading a tag using the electrical rotation principle based on two discrete transmitting angles. a) Tag measurement results at $\alpha_t$ = 0°, and the tag position corresponds to $\beta$ = 0° and 45°. b) Tag measurement result at $\alpha_t$ = 45°. For a color version of this figure, see www.iste.co.uk/garbati/chipless.zip*

Figure 6.14 shows the normalized measurement using the STFT of a cross-polarized tag, for different tag positions, with a fixed $\alpha_t$ = 0°. Note that to ease the comparison, all the signals have been normalized. Therefore, in Figure 6.14, we present the dependence of RCS amplitude on the orientation that can be retrieved from Figure 6.12. The test bench is the one shown in Figure 6.10(a). The tag position β varied between 0°, 15°, 30° and 45°. The reading was successful when $\beta - \alpha_t$ was within 30°. It corresponds to the case of $\beta$ = 0°, 15° and 30°. As expected, the best result is for $\beta = \alpha_t$ ($\beta - \alpha_t = 0$), which corresponds to one maximum of equation [6.12]. This result validates the assumption that the tag can be read despite not being perfectly oriented with the antenna of the reader, i.e. within a certain range around the maximum of equation [6.12]. In this case, it was read at an angle of 30° from the maximum.

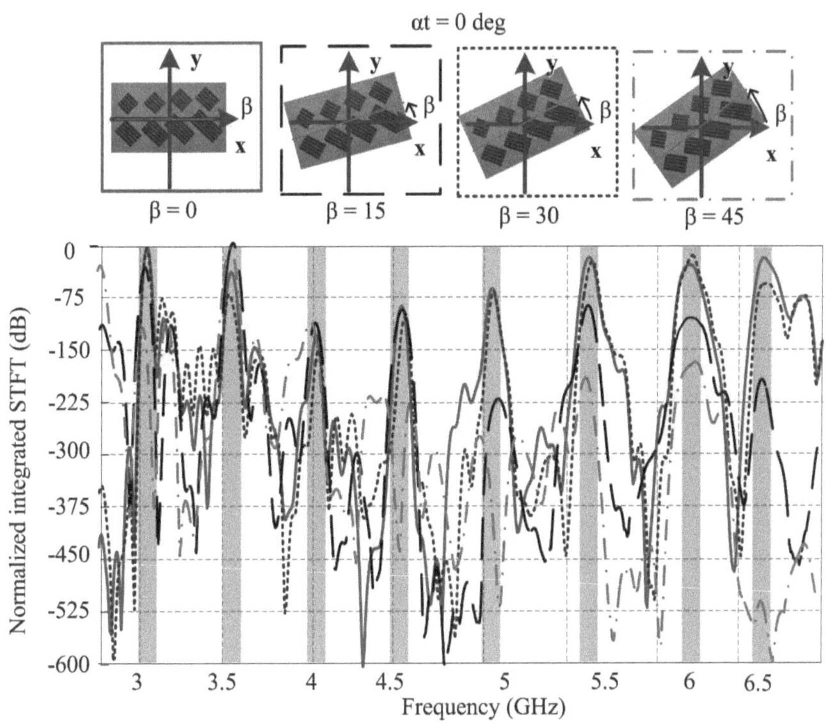


**Figure 6.14.** *Normalized measurement using the STFT of a cross-polarized tag, for different tag positions with a fixed $\alpha_t$ = 0°. The bench is the one presented in Figure 6.10(a). The tag position $\beta$ varies between 0°, 15°, 30° and 45°. The reading was successful for all values of $\beta$, except for 45° (minimum of equation [6.12]). For a color version of this figure, see www.iste.co.uk/garbati/chipless.zip*

The two measurements are required to read the tag independently of its orientation, as demonstrated in Figures 6.13 and 6.14. In the following section, this approach will be applied to the optimized reader described in Chapter 4. A criterion of selection is used to select the correct measurement approach to retrieve the tag ID.

## 6.5. Optimized reader approach

The chipless reader version, presented in Chapter 4, may be modified to suit the tag of the linearly polarized interrogating signal as for the VNA. It must perform two measurements with $\alpha_t$ varying by 45°, according to the result of the previous section. The proposed reader is shown in Figure 6.15. It is composed of a central block with some additional components. The transmission part is composed of the pulse generator HyperLabs HL9200, followed by a wideband splitter ZN2PD-63+ from Mini-Circuits. One of its outputs is directly connected to the port, allowing horizontal polarization of the Satimo QH2000 antenna. The other splitter output is connected to a wideband switch, with one output closed to a 50 Ω load, and the other to the second port of the same transmitting antenna, allowing vertical polarization. In reception, another dual polarized antenna is connected to a switch and then to the HPF Mini-Circuit VHF-3100+ and the LNA Mini-Circuits TVA-82–213A+. Both the transmission and reception chains are connected to the optimized reader presented in Chapter 4, which is composed of the Xilinx FPGA evaluation board SP601, the designed RF-front end and the alimentation board.

In transmission, the reader is able to send a linearly polarized signal with $\alpha_T$ = 0° and 45° for the configurations *Conf. 1* and *Conf. 2* (see Figure 6.15). In reception, by changing the position of the switch between $R_A$ and $R_B$, the tag response in two orthogonal directions can in turn be measured. Finally, in post-processing, it is possible to isolate the cross-component depending on the imposed transmitting angle $\alpha_t$. For $\alpha_t$ = 0° (*Conf. 1*), the reader must measure only the vertical contribution of tag response, as the switch in the reception chain is set in the position $R_B$. For $\alpha_T$ = 45° (*Conf. 2*), both orthogonal components need to be measured (position $R_A$ and thereafter $R_B$). Then, to isolate the electrical-field contribution in the cross-direction ($\alpha_t$ + 90°), as described in equation [6.9], it is sufficient to subtract the horizontal contribution from the vertical.

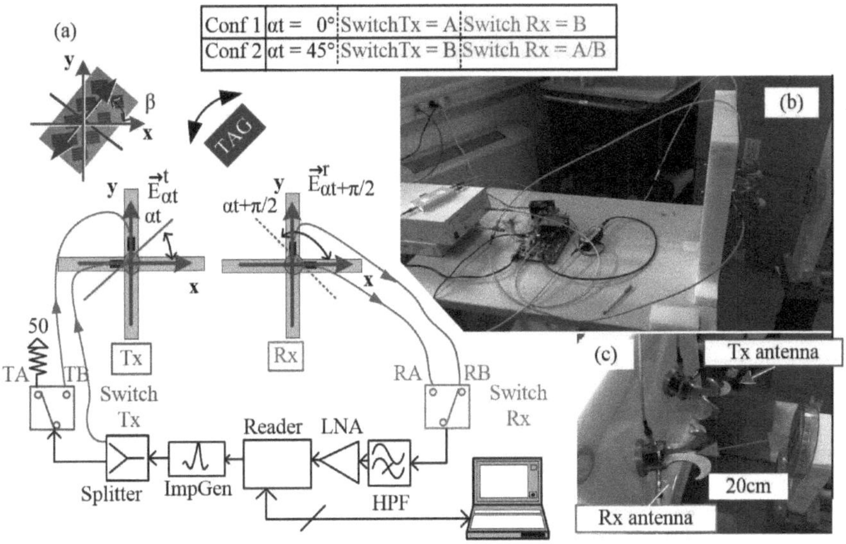


**Figure 6.15.** *Modification of the optimized reader presented in Chapter 4 to perform the measurement with $\alpha_t$ in two discrete positions (0° and 45°)and to retrieve the tag ID independently of its orientation. a) Reader architecture, b) measurement bench in a real environment and c) photograph of antennas and tag positions. For a color version of this figure, see www.iste.co.uk/garbati/chipless.zip*

The reader has an external LNA to compensate for multiple attenuating factors. First, the use of a splitter in transmission reduces the transmitted power to a value as low as 3 dB in the case of Conf. 1 ($\alpha_t = 0°$). For Conf. 2 ($\alpha_t = 45°$), limitations are in reception because the tag response in the cross-component ($\alpha_t + 90°$) with $\alpha_t = 45°$ is in turn measured through its projections along with the two receiving antenna components and therefore attenuated of 3 dB with respect to the main signal. Thus, the effective ADC number of bits used may decrease, as shown in Chapter 4 (section 4.2), with increasing quantization noise. The second attenuating factor is the use of two distinct dual-access antennas instead of only one, similar to the classical reading technique. With only one antenna, the tag can be well aligned, whereas with two antennas, it is more difficult and less effective, especially when the distance of the tag varies. Moreover, the antennas are designed to operate from 2 GHz and therefore are larger than required for UWB applications (3.1–10.6 GHz).

Figure 6.16 shows measurements performed in a practical environment with the reader in *Conf. 1* ($\alpha_t = 0°$), where the tag position β is gradually varied from the best case ($\beta = 0°$) to the worst case ($\beta = 45°$). The setup is shown in Figure 6.15(b), where the tag was placed at a distance of 20 cm from the antennas, as shown in Figure 6.15(c). First, an empty measurement was performed without the tag, and then the second measurement was performed with the tag. As predicted, the reader was able to read the tag at $\beta$ ranging from 0° to 30°. At 45°, the reading was incorrect because the useful signal level was excessively low. To read the tag in this last orientation, *Conf. 2* must be used ($\alpha_t = 45°$).

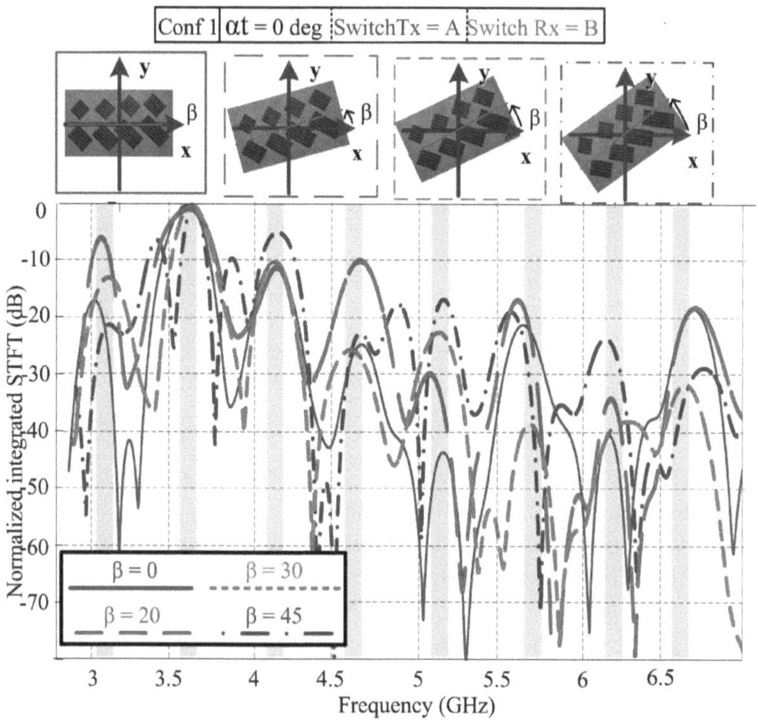


**Figure 6.16.** *Normalized measurement results of the bench presented in Figure 6.15(b) in a practical environment. The tag was oriented in different directions, from $\beta$ = 0–45° using Conf. 1 ($\alpha_t$ = 0°). For a color version of this figure, see www.iste.co.uk/garbati/chipless.zip*

Figure 6.17 shows the measurement of a tag placed at 45° ($\beta$) using the two configurations *Conf. 1* (blue dashed line) and *Conf. 2* (red straight line), respectively. As expected, *Conf. 2*, which matches the position of the tag, shows the best reading ($\alpha_t = 45°$). In *Conf. 1,* only the first resonance is in the correct position, whereas in *Conf. 2,* almost all the resonance positions are correctly placed. Moreover, the second resonance (and, to a lesser extent, the third resonance) is shifted by a few tens of MHz *in Conf. 2*. This measurement deviation is due to many factors which could be improved. In addition to those already mentioned, important contributions come from the adaptation of antennas, which is above –10 dB up to 4 GHz. Finally, the reader prototype is composed of numerous distinct devices, some of which are on the FR-4 substrate.

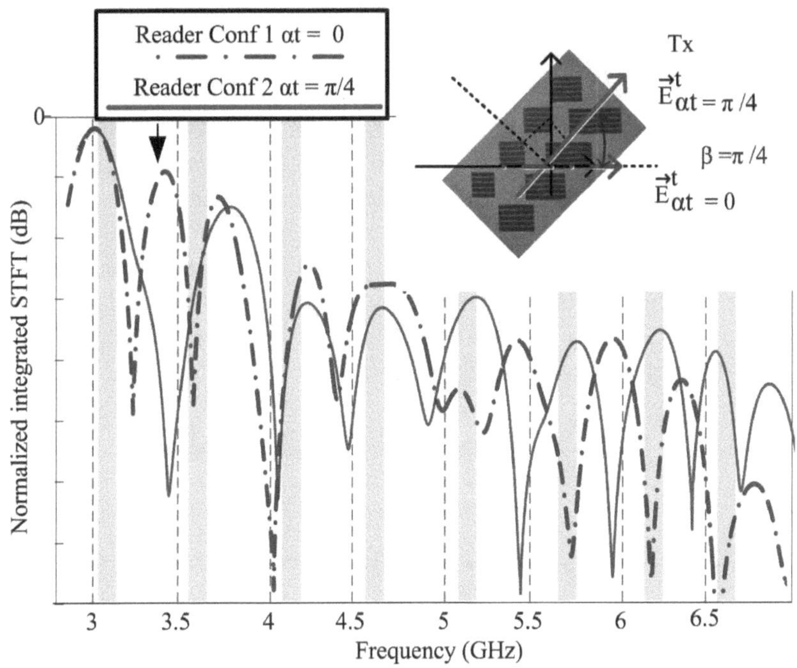


**Figure 6.17.** *Tag placed at β = 45° and read with both reader configurations. The measurement was executed in a practical environment and the setup is shown in Figure 6.15(b). For a color version of this figure, see www.iste.co.uk/garbati/ chipless.zip*

As demonstrated, the reading in both configurations (*Conf. 1* and *Conf. 2*) may guarantee a correct reading of the tag independent of the tag orientation. However, a criterion is needed to select the correct measured response. From Figure 6.16, in some cases, it is possible to discern the correct reading between the two configurations only if the expected number of peaks (resonant frequencies) for a given frequency range is already known. In a more general case, a criterion of selection is needed. We may keep the last peak position fixed as a criterion and use the others to encode the tag ID. However, this will reduce the tag coding capacity, and will not prevent the situation where both configurations (Conf. 1 and Conf. 2) give a correct position of the last peak. The suggested approach is to distinguish the two signals based on the maximum tag signal power. Figure 6.17 shows the response of the tag shown in Figure 6.16 in the time domain. For clarity, the signal in the time domain has been cut at 8 ns. The tag structural mode prevails in the first 2 ns, followed by the antenna (tag) mode, where the signal contains the coding information. The tag structural mode was removed by using the STFT. The power of the structural mode in *Conf. 2* (correct reading) is higher than that in *Conf. 1*. However, the use of this power difference as a criterion is considered to be a risk because it can vary based on the item to which the tag is attached. Thus, a discrimination based on the antenna (tag) mode is more appropriate. The selected measurement to be used to decode the tag ID will be the more powerful one inside the tag mode frame [REZ 15]. This algorithm frees the tag designer from any additional constraints from the reader part.

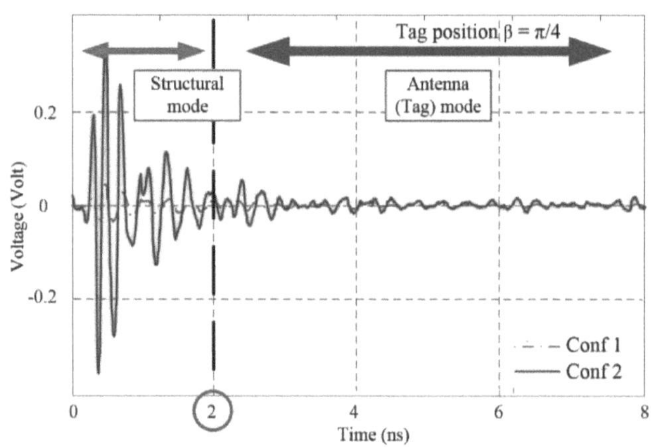


**Figure 6.18.** *Response of the measurement shown in Figure 6.17 in the time domain. The tag was placed at β = 45° and interrogated with both reader configurations (Conf. 1 and Conf. 2). For a color version of this figure, see www.iste.co.uk/garbati/chipless.zip*

## 6.6. Conclusion

This chapter proposed a reading system for UWB chipless RFID technology, which is independent of tag orientation. It is an original approach because, in contrast to the other RFID systems, it is not based on circular polarized antennas, and the reading is specific to chipless tags that are designed based on the cross-polarization principle.

The concept for electrically changing the reader antenna polarization has been presented and validated in an anechoic environment. Measurements of a chipless tag at different polarization orientations show that the tag can be read with a fixed antenna while keeping the cross-polarization configuration. The first reader system was based on a state-of-the-art four-port VNA, and its ability to perform balance measurements was exploited. Then, a second reader was realized based on the hardware implementation proposed in Chapter 4. It is able to perform measurements at two discreet positions, which are sufficient to correctly retrieve the tag ID. A criterion to discriminate the correct measurement approach to recover the ID was introduced. This reader does not impose any constraint on the designer with respect to the tag shape or even to the encoding approach to be used. In order to improve the reader performance, dedicated work on the antennas part is required, with the aim of reducing their dimensions in order to be better aligned with the tag. Furthermore, future work is expected to include component integration and the use of an ADC with a better resolution.

# Appendices

# Appendix 1

## Matlab® GUI Acquisition Software for Agilent DSO91204A

To facilitate the use of the high-performance DSO Agilent 91204A, a GUI has been developed in the Matlab® environment. It was used to debug all the designed electronics cards presented in this book, as well as to read UWB chipless tags in an IR-UWB configuration.

A screenshot of the Matlab® GUI is shown in Figure A.1. It is divided into four rows, each representing a different acquisition channel of the DSO but sharing the same characteristics. A radio button allows us to switch between measurements with and without trigger. A measurement without trigger means that the GUI transfers the actual signal on the memory of the DSO to the PC for a particular channel. If the measurement with trigger is selected, the DSO will start a new acquisition and transfer the data of the selected channel to the PC. Finally, a button allows us to simultaneously collect the data of all the channels.

To communicate with the DSO using the Matlab® environment, the instrument control toolbox must be installed on the PC. It contains a library of functions that can make use of the Agilent IO Libraries Suite. The latter is a collection of libraries and software utilities to enable the communication with the DSO. It also contains the virtual instrument software architecture (VISA) libraries that are used by the developed Matlab® GUI.

To establish communication with the DSO, the following lines of code are used:

```
% Create the Visa object v for the DSO using the USB interface
v=visa('agilent', 'USB0::0x0957::0x9002::MY51050102::0::INSTR');
% Remove data from the input buffer of the object v
flushinput(v);
% Allocate the object v input buffer
v.InputBufferSize=buffersize;
% Opening the communication with the DSO
fopen(v);
% Define a time out of 10 seconds
v.Timeout=10;
```

To query for the data of a precise channel with trigger, the next lines of code are used. If the measurement is without trigger, it is sufficient to comment the first line of this code:

```
% Single trigger section
fprintf(v,':SINGLE');
% Select the Channel Chan
fprintf(v,':WAVEFORM:SOURCE CHAN');
% Define the waveform format in byte
fprintf(v,':WAVEFORM:FORMAT BYTE');
% Query for channel vertical increment and origin
yinc=str2double(query(v,':WAVEFORM:YINC?'));
yor=str2double(query(v,':WAVEFORM:YOR?'));
% Query for channel horizontal (time) increment and origin
xinc=str2double(query(v,':WAVEFORM:XINC?'));
xor=str2double(query(v,':WAVEFORM:XOR?'));
% Query for channel measurement
fprintf(v,':WAVEFORM:DATA?')
% Save measurement in variable y
y = binblockread(v,'int8');
fread(v);
```

The following lines of code can be used to close the communication:

```
% Clear object v
clrdevice(v);
% Close the communication with object v
fclose(v);
% Remove object v
delete(v);
```

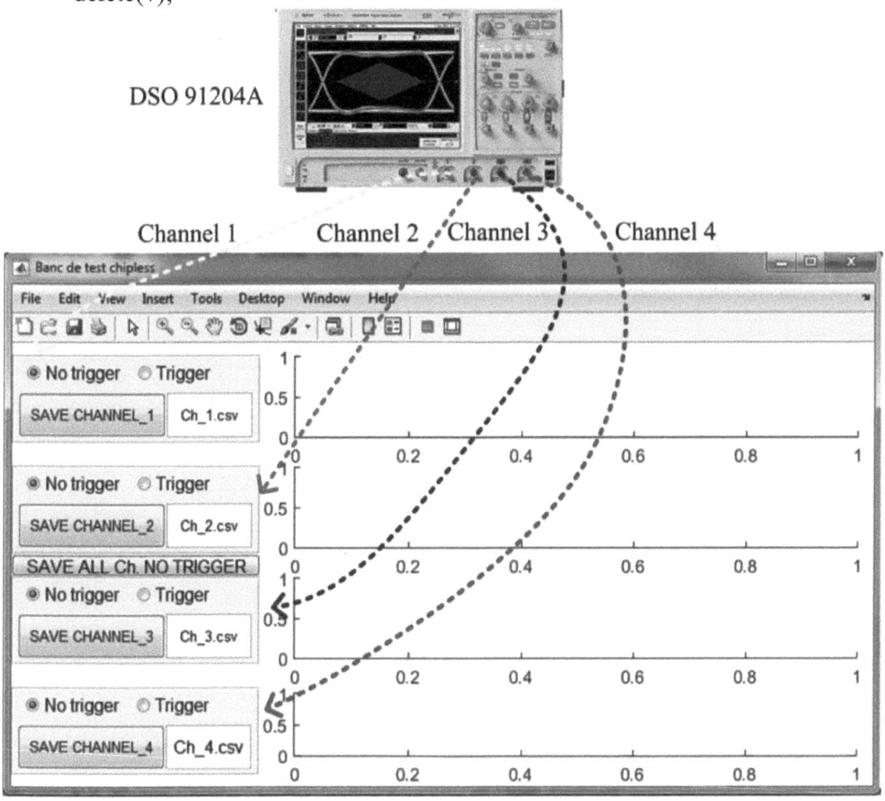


**Figure A1.1.** *Matlab® GUI designed to acquire the Agilent 91204A DSO measuring data. It can save the data from all the four channels of the DSO. For a color version of this figure, see www.iste.co.uk/garbati/chipless.zip*

# Appendix 2

# PC Application Software for First Reader Version in Chapter 3

The IR-UWB reader version designed during the project Gravit came with a dedicated application software for the PC. It was written in Matlab® and is based on different scripts that are used by the GUI, as shown in Figure A2.1. It expects the execution of a tag measurement according to two simple measurements, namely the background subtraction measurement (without a calibrating tag) and tag measurement. The GUI was designed to read frequency-coded tags, presented in [VEN 13a], which are based on coupled short dipoles with the REP and cross-polarization approaches. They are presented in section 1.3 of Chapter 1 and used throughout this book to characterize the performance of the designed readers. A photograph of the tag is also shown in Figure A2.1.

The GUI has two main columns: the first controls the acquisition from the reader and the second is in charge of post-processing and retrieving tag ID. On the top left, there is an option to open the communication with the COM port of the reader and to select the averaging factor. Once acquisition with the *Measure* command is started, the reader executes the reading process and transfers the results to the GUI, which will show it in the time domain in the corresponding first window.

The *Files* section is designed to memorize the acquired measurement on clicking the relative *Save* button. The GUI requires two measurements in the database, namely the tag measurement and the background measurement to perform the post-processing.

The RCS calculator section defines the timing window inside the frame of acquisition of 120 ns, where tag decoding should be executed. The identifier section offers the possibility of correlating the response of the tag with the response of an ideal resonator, which is emulated in amplitude with a Gaussian function. Finally, at the bottom, once the resonant frequencies of the measured tag are retrieved, their position is compared with a database, and the tag ID is generated. According to section 1.3 of Chapter 1, the backscattering measurement is only used to characterize the coupling of the reader antennas (leakage). It is convenient to perform this calibration only once in a practical application. Therefore, the GUI is able to automatize the reading process in a loop execution of tag reading with the button *Read* and by selecting the checkbox *Loop*.

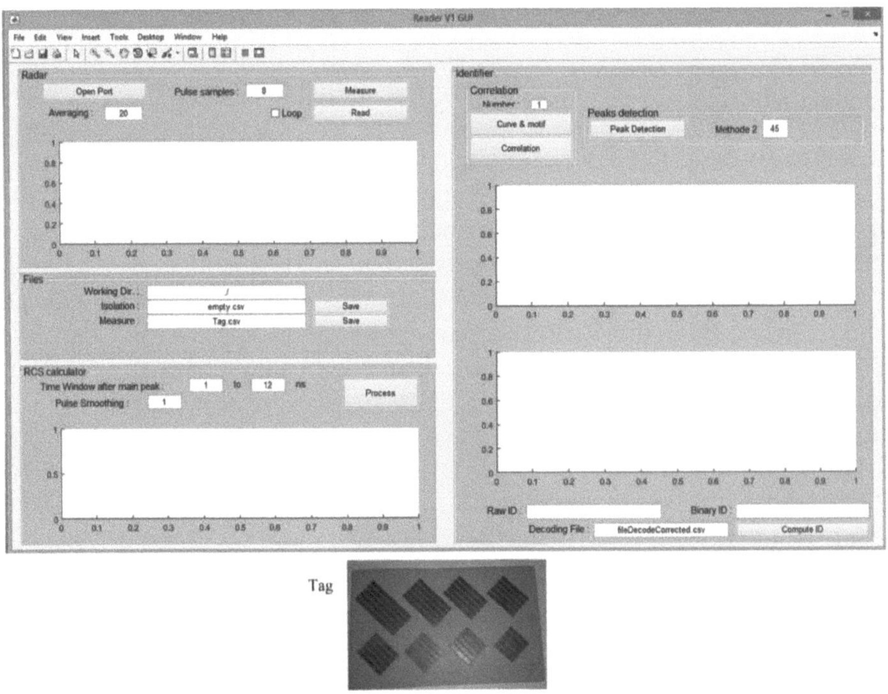


**Figure A2.1.** *GUI designed in Matlab® to read the tags shown in the figure, with the reader designed during the project Gravit. The tag was of the same family presented in [VEN 13a] and used throughout this book to characterize the performance of the designed readers*

## A2.1. PC–reader communication

The reader communicates with the PC using a USB port, where a UART is emulated. Inside the Xilinx SP601 FPGA evaluation board, there is a CP2103GM chip, which is a USB-to-UART chip, from Silicon Labs. It is connected to the FPGA from the UART side and to the PC from the USB side. The Matlab® GUI uses the Windows COM library to communicate with the established COM connection with the reader. The GUI creates a serial port object, called *s*, with the following function:

```
s = serial(port,'BaudRate', 460800,'TimeOut', 80,'InputBuffer', 16384);
```

The BR is 460.800 KB and the serial time-out is 80 s, which covers the maximum time required by the reader to complete an acquisition on receiving the command from the PC. For the maximum averaging of 128, it is about 64 s.

The Matlab® function in charge of communicating with the reader is called getRadarFrame(~,~). The chunk of code that queries the reader for a reading process and retrieves the reader for measured data is as follows:

```
% Get the averaging factor written in the corresponding editing box
avg = str2double(get(data.RadarAvg,'String'));
% Convert the averaging factor from double to string
str = sprintf('G%03d', avg);
% Transmit the command to the reader
fprintf(s, str);
% Receiving the information about the number of acquired points from the
reader
numSamples = str2double(fgetl(s));
% Reading from the reader
Tampon = zeros(1, numSamples);
for i = 1 : numSamples
        0054ampon(i) = str2double(fgetl(s))-512;
end
```

# Appendix 3

## Matlab® GUI Reader in Chapter 4

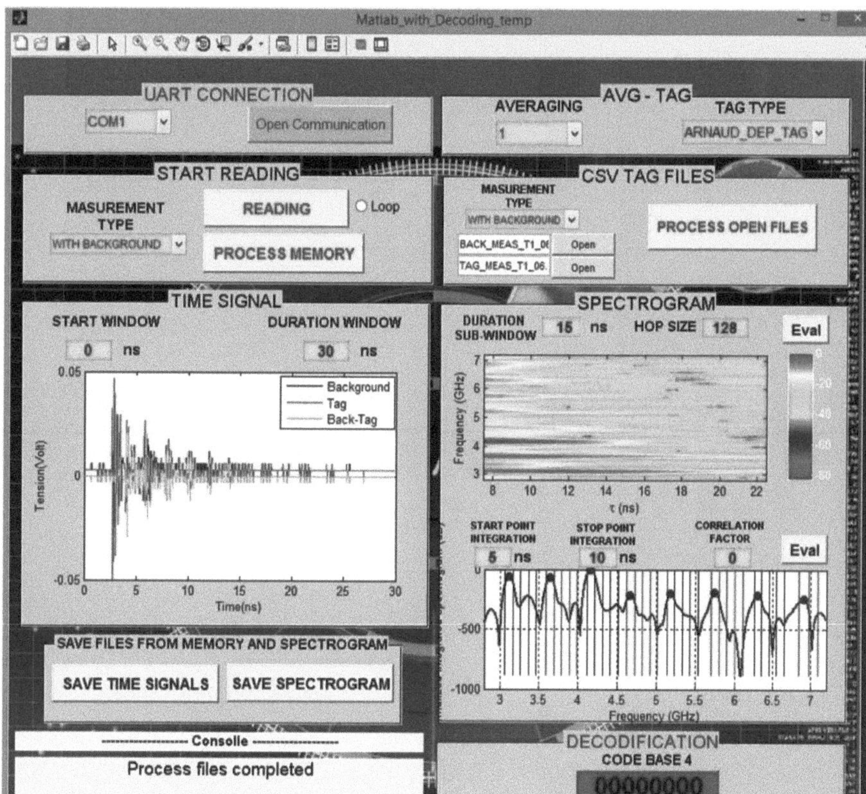


**Figure A3.1.** *Matlab® GUI Reader presented in Chapter 4. With respect to the GUI for the reader version developed during the project Gravit, it introduces the STFT analysis. For a color version of this figure, see www.iste.co.uk/garbati/chipless.zip*

The Matlab® GUI for the reader proposed in Chapter 4 is shown in Figure A3.1. It presents the same tag post-processing treatment of the GUI for the reader presented in Chapter 3, with the addition of the STFT shown with spectrogram utility. The STFT for REP-based UWB frequency-coded chipless RFID tags was studied in [RAM 16b]. According to [RAM 16b], the spectrogram was integrated in the time domain between *start point integration* and *stop point integration.* The STFT window was rectangular, and its duration settled with *DURATION SUB-WINDOW.* As a result of the reader performance and post-processing STFT implementation, the background measurement may sometimes be omitted, and therefore two pop-up menus of the Matlab® GUI allow its selection.

# Appendix 4

## Schematic Power Supply Board Reader in Chapter 4

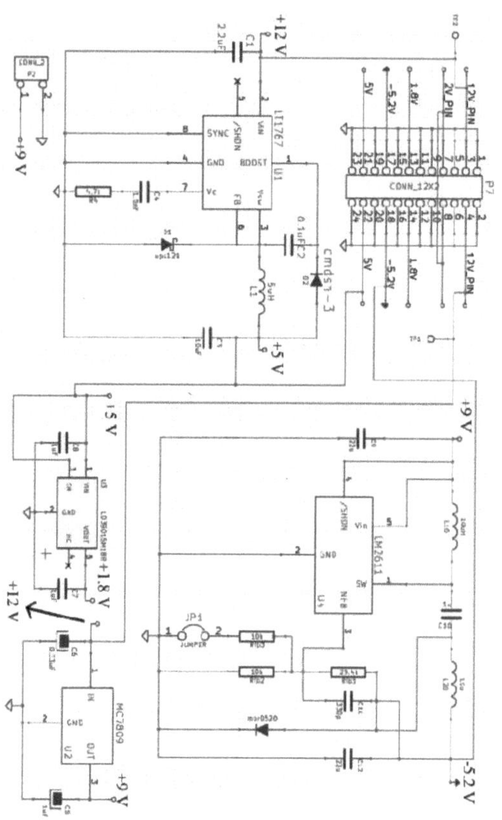


**Figure A4.1.** *Schematic of the power supply board for the reader presented in Chapter 4. For a color version of this figure, see www.iste.co.uk/garbati/chipless.zip*

Figure A4.1 shows the schematic of the power supply board designed for the reader presented in Chapter 4. It generates the required voltage values for the RF front-end from the two 12 V pins available on the Xilinx FPGA evaluation board SP601.

# Appendix 5

## Matlab® GUI Acquisition Software for VNA N5222A Performing Balanced Measurement

To simplify the use of the VNA N5222A, when performing a balanced measurement, a user-friendly GUI in Matlab® has been developed. It is interactive and changes items' visibility based on the current connection state. A console is used to inform the user about the measurement evolution. The GUI at the software initialization is shown in Figure A5.1(a). It expects the selection of one or both balanced measurements Scd21 and Scc21, in true mode or not. After the communication of software with the VNA, the GUI changes its appearance, as shown in Figure A5.1(b).

The power and/or the phase between the signals of the two physical ports 2 and 4 of logic port 2 can be varied. The power variation range is retained within ±15 dB to avoid a possible VNA error message. The option *offset as fixture* can also be selected. Varying the power offset in true mode corresponds to a variation of $\alpha_t$ in the range of [10°–80°], as shown in Table 6.1 of Chapter 6. Finally, the GUI allows an internal averaging of the measured results besides that of the VNA. It performs several queries of the measurement process of the VNA.

To use the Matlab® environment to communicate with the VNA, the instrument control toolbox must be installed on the PC. It contains a library of functions that can exploit the Agilent IO Libraries Suite, which is a collection of libraries and utility software to enable the communication with the instrument. It also contains the VISA libraries that are used by the developed Matlab® GUI.

To establish communication with the VNA, the following lines of code are used:

```
% Create the Visa object obj1 for the VNA, using the USB interface
obj1 = visa('AGILENT', 'USB0::0x0957::0x0118::MY51421171::0::INSTR');
% Set input buffer size for largest possible read of 32001 points in float32
(32001*4*2)
set(obj1,'InputBufferSize', 260000);
% Opening the communication with the VNA
fopen(obj1);
% Active continuously trigger system
fwrite(obj1, 'INIT:CONT ON');
% Single precision floating points (32 bit)
fwrite(obj1, 'FORM REAL, 32');
% Byte orders (LSB first – little-endian)
fwrite(obj1, 'FORM:BORD SWAP');
%Delete all measurements on VNA (clean the VNA)
fprintf(obj1, 'CALC:PAR:DEL:ALL');
```

To set up a differential measurement, the following code is used, which refers to the case of Scc21; for Sdc21, a similar code can be used:

```
%Define one measurement
fwrite(obj1,'CALC:PAR:DEF:EXT "MYSCC12","S11"');
%Select the measurement just created
fwrite(obj1,'CALC:PAR:SEL "MYSCC12"');
% Set balanced parameters ON
fwrite(obj1,'CALC:FSIM:BAL:PAR:STAT ON');
% Set balanced transformation to BBAL (balance to balance)
fwrite(obj1,'CALC:FSIM:BAL:DEV BBAL');
% Allow for variable phase offset between physical ports of the same logic
port
fwrite(obj1,'CALC:FSIM:BAL:FIXT:OFFS:PHAS OFF');
% Allow for variable power offset between physical ports of the same logic
port
fwrite(obj1,'CALC:FSIM:BAL:FIXT:OFFS:POW OFF');
% Define balanced measurement to SCC12
```

```
fwrite(obj1,'CALC:FSIM:BAL:PAR:BBAL:DEF SCC12');
% Display trace
fwrite(obj1,'DISP:WIND1:TRAC1:FEED "MYSCC12"');
if(TRUEMODE)
    % Enable true mode measurement
    fwrite(obj1,'CALC:FSIM:BAL:STIM:MODE TM');
else
    fwrite(obj1,'CALC:FSIM:BAL:STIM:MODE SE'); % Single-Ended
    stimulus
end;
```

To vary the power offset or the phase offset on the balance port 2:

```
fprintf(obj1,'CALC:FSIM:BAL:BPOR2:OFFS:PHAS %g',phase_i);
fprintf(obj1,'CALC:FSIM:BAL:BPOR2:OFFS:POW %g',power_t);
```

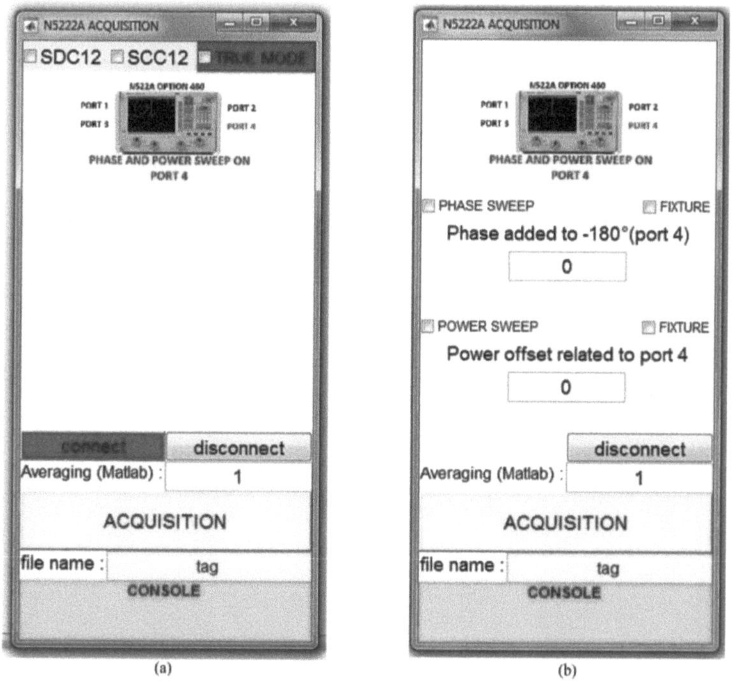


**Figure A5.1.** *Matlab® GUI designed to acquire the Agilent N5222A VNA measuring data in balanced mode. a) GUI that appears at software initialization before selecting the desired measurements (Sdc12-Scc12) and before being connected to the VNA. b) GUI after the selection of measurement and connection. It is interactive and changes items' visibility based on the current connection state. A console is used to inform the user about the measurement evolution*

# Bibliography


[ANA 17] ANALOG DEVICES, ADA4930-1 data sheet, available online at: www.analog.com, 2017.

[BAU 91] BAUM C.E., ROTHWELL E.J., CHEN K. *et al.*, "The singularity expansion method and its application to target identification", *Proceedings of the IEEE*, vol. 79, no. 10, pp. 1481–1492, 1991.

[BEN 48] BENNETT W.R., "Spectra of quantized signals", *The Bell System Technical Journal*, vol. 27, no. 3, pp. 1–3, July 1948.

[BJO 16] BJORNIEN T., SUIKKOLA J., UKKONEN L. *et al.*, "Testing and modeling the performance of stretchable screen printed UHF RFID tag under strain", *2016 IEEE International Symposium on Antennas and Propagation (APSURSI)*, pp. 465–466, Fajardo, 2016.

[CHA 00] CHANG K., "Radar and Sensor Systems", *RF and Microwave Wireless Systems*, John Wiley & Sons, 2000.

[DEA 07] DE ANGELIS A., DIONIGI M., MOSCHITTA A. *et al.*, "A low-cost ultra-wideband indoor ranging technique", *IEEE Instrument and Measurement Technology Conference IMTC 2017*, pp. 1–6, 2007.

[DEA 08] DE ANGELIS A., DIONIGI M., GIGLIETTI R. *et al.*, "Experimental low-cost short pulse generators", *IEEE Instrument and Measurement Technology Conference*, pp. 259–564, 2008.

[DEL 98] DELIYANNIS T., YICHUANG S., FIDLER J.K., *Continuous-Time Active Filter Design*, CRC Press, Boca Raton, 1998.

[DOB 05] DOBKIN D.M., "A radio-oriented introduction to RFID – protocols, tags and application", *High Frequency Electronics*, August 2005.

[DOB 12] DOBKIN D.M., *The RF in RFID: UHF RFID in Practice*, 2nd Edition, Newnes Newton, 2012.

[ETS 04] EUROPEAN TELECOMMUNICATION INSTITUTE (ETSI), Electromagnetic compatibility and Radio spectrum Matters (ERM); Short Range Devices (SRD); Technical characteristics for SRD equipment using Ultra Wide Band technology (UWB) Part 1: Communications applications, ETSI TR 101 994-1 V1.1.1, 2004.

[ETS 10] EUROPEAN TELECOMMUNICATION INSTITUTE (ETSI), Electromagnetic compatibility and Radio spectrum Matters (ERM); Short Range Devices (SRD) using Ultra Wide Band technology (UWB) for communications purposes; Harmonized EN covering the essential requirements of article 3.2 of the RTTE Directive, ETSI EN 302 065 V1.2.1, 2010.

[ETS 17] EUROPEAN TELECOMMUNICATION INSTITUTE (ETSI), Electromagnetic compatibility and Radio spectrum Matters (ERM); Short Range Devices (SRD) using Ultra Wide Band technology (UWB) for communication purposes; harmonized EN covering the essential requirements of article 3.2 of the R&TTE Directive, ETSI E. 302 065 V1.2.1, available online at: http://www.etsi.org/, 2017.

[FCC 17] FCC, Title 47, Telecommunication, Chapter 1, subchapter A, Part 15, subpart F – Ultra-wideband operation, available online at: https://www.ecfr.gov/cgi-bin/text-idx?SID=9f76f123f0f69eb624596f8d8fd06ce6&mc=true&node=sp47.1.15.f&rgn=div6.

[FEN 15] FENG Y., XIE L., CHEN Q. *et al.*, "Low-cost printed chipless RFID humidity sensor tag for intelligent packaging", *IEEE Sensors Journal*, vol. 15, no. 6, pp. 3201–3208, June 2015.

[FLE 02] FLETCHER R.R., Low-cost electromagnetic tagging: design and implementation, PhD thesis, Massachusetts Institute of Technology, Cambridge, 2002.

[FON 04] FONTANA R.J., "Recent system applications of short-pulse ultra-wideband (UWB) technology", *IEEE Transaction on Microwave Theory and Technique*, vol. 54, no. 9, pp. 2087–2104, 2004.

[GAR 15] GARBATI M., SIRAGUSA R., PERRET E. *et al.*, "Low cost low sampling noise UWB Chipless RFID reader", *IEEE MTT-S International Microwave Symposium (IMS)*, pp. 1–4, 2015.

[GEN 08] GENG Z., SIMROCK S.N., "Evaluation of fast ADCs for direct sampling RF field detection for the European XFEL and ILC", *24th Linear Accelerator Conference LINAC08*, Victoria, September 2008.

[GHA 07] GHAVAMI M., MICHAEL L.B., KOHNO R., *Ultra-Wideband Signals and Systems in Communication Engineering*, John Wiley & Sons, 2007.

[GIR 13] GIRBAU D., RAMOS A., LÁZARO A. *et. al.*, "UWB time-coded RFID sensors: a comparison between passive and semi-passive approaches", *ICECom*, pp. 1–4, Dubrovnik, 2013.

[GRE 63] GREEN R.B., The general theory of antenna scattering, PhD thesis, Ohio State University, 1963.

[HAM 12] HAMDI M., GARET F., DUVILLARET L.P. *et al.*, "New approach for chipless and low cost identification tag in the THz frequency domain", *2012 IEEE International Conference on RFID-Technologies and Applications (RFID-TA)*, pp. 24–28, Nice, 2012.

[HAR 02] HARTMANN C.S., "A global SAW ID tag with large data capacity", *2002 IEEE Ultrasonics Symposium Proceedings*, vol. 1, pp. 65–69, 2002.

[HEW 84] HEWLETT PACKARD, Pulse and waveform generation with step recovery diodes, Application Notes AN918, 1984.

[JAL 05] JALALY I., "RF barcodes using multiple frequency bands", *IEEE MTT-S International Microwave Symposium Digest (IMS)*, pp. 139–142, 2005.

[KAH 15] KAHAR M., MANDAL J., MANDAL M.K., "A compact Vivaldi antenna with microstrip feed line for the 2–18 GHz ultra-wideband applications", *2015 IEEE Applied Electromagnetics Conference (AEMC)*, pp. 1–2, Guwahati, 2015.

[KEY 16a] KEYSIGHT TECHNOLOGIES, Spectrum Analysis Basics, Application Note 150, 2016.

[KEY 16b] KEYSIGHT TECHNOLOGIES, PNA Series Network Analyzers Help, available online at: http://na.support.keysight.com, 2016.

[KHE 14] KHEIRDOOST A., MORADI G., ELKHOLY E. *et al.*, "Modeling and jitter improvement of SRD-based ultra-wideband pulse generator", *IEEE Transaction on Microwave Theory and Technique*, vol. 62, no. 8, pp. 1736–1747, August 2014.

[KIM 08] KIM J., Saw based chipless passive RFID tag using cellulose paper as substrate and method for manufacturing the cellulose paper, Patent WO/2008/056,848, 2008.

[KOS 12] KOSWATTA R.V., KARMAKAR N.C., "A novel reader architecture vased on UWB chirp signal interrogation for multiresonator-based chipless RFID tag reading", *IEEE Transaction on Microwave Theory and Technique*, vol. 60, no. 9, pp. 2925–2933, September 2012.

[LAZ 11] LAZARO A., RAMOS A., GIRBAU D. *et al.*, "Chipless UWB RFID tag detection using continuous wavelet transform", *IEEE Antennas Wireless Propagation Letters*, vol. 10, pp. 520–523, 2011.

[LE 08] LE T., MAYARAM K., FIEZ T., "Efficient far-field radio frequency energy harvesting for passively powered sensor networks", *IEEE Journal of Solid-State Circuits*, vol. 43, no. 5, pp. 1287–1302, May 2008.

[LEE 14] LEE K.K., LANDE T.S., "A wireless-powered IR-UWB transmitter for long-range passive RFID tags in 90-nm CMOS", *IEEE Transaction on Circuits Systems II*, vol. 61, no. 11, pp. 870–874, 2014.

[LI 12] LI Q., JIANG W., XU Y. *et al.*, "Analysis and design of wide-band comb generator based on SRD", *IEEE International Conference on Material and Manufacturing Technology (ICMMT)*, pp. 1–2, 2012.

[LIU 13] LIU Q., WANG Y., FATHY A.E., "Towards low cost, high speed data sampling module for multifunctional real-time UWB radar", *IEEE Transaction on Aerospace and Electronic Systems*, vol. 49, pp. 1301–1316, 2013.

[MEN 05] MENZEL W., RAHMAN TITO M.S., ZHU L., "Low-loss ultra-wideband (UWB) filters using suspended stripline", *IEEE Microwave Conference Proceedings*, 2005.

[MIC 02] MICROCHIP, microID 125 kHz RFID System Design Guide, Application Note AN680, 2004.

[MIT 09] MITZNER K., Complete PCB Design Using OrCAD Capture and PCB Editor, Newnes, 2009.

[MOO 12] MOORFELD R., LU Y., FINGER A., "Energy detection with optimal symbol constellation for M-PAM in UWB fading channels", *IEEE International Conference on Ultra-Wideband* (ICUWB), vol. 54, pp. 397–401, 2012.

[NAI 16] NAIR R.S., PERRET E., "Folded multilayer C-sections with large group delay swing for passive chipless RFID applications", *IEEE Transactions on Microwave Theory and Techniques*, vol. 64, no. 12, pp. 4298–4311, December 2016.

[NIJ 12] NIJAS C.M., DINESH R., DEPPAK D. *et al.*, "Chipless RFID tag using multiple microstrip open stub resonators", *IEEE Transactions on Antennas and Propagation*, vol. 60, no. 9, pp. 4429–4432, September 2012.

[NXP 17] NXP, M68HC11 Reference Manual, available online at: https://www.nxp.com, 2017.

[OH 11] OH S., WENTZLOFF D.D., "A step recovery diode based UWB transmitter for low-cost impulse generation", *IEEE International Conference on Ultra-Wideband (ICUWB)*, pp. 63–67, 2011.

[PER 14] PERRET E., *Radio Frequency Identification and Sensors: From RFID to Chipless RFID*, ISTE Ltd, London and John Wiley & Sons, New York, 2014.

[PET 05] PETTUS M., RFID system utilizing parametric reflective technology, Patent US20050280539 A1, 2005.

[POP 16a] POPPERL M., ADAMETZ J., VOSSIEK M., "Polarimetric radar barcode: a novel chipless RFID concept with high data capacity and ultimate tag robustness", *IEEE Transactions on Microwave Theory and Techniques*, vol. 64, no. 11, pp. 3686–3694, November 2016.

[POP 16b] POPPERL M., CARLOWITZ C., VOSSIEK M. *et. al.*, "An ultra-wideband time domain reflectometry chipless RFID system with higher order modulation schemes", *German Microwave Conference (GeMiC)*, pp. 401–404, 2016.

[PRE 09] PRERADOVIC S., BALBIN I., KARMAKAR N.C. *et al.*, "Multiresonator-based chipless RFID system for low-cost item tracking", *IEEE Transactions on Microwave Theory and Techniques*, vol. 57, no. 5, pp. 1411–1419, May 2009.

[PRE 10a] PRERADOVIC S., KARMAKAR N.C., "Chipless RFID: Bar Code of the Future", *IEEE Microwave Magazine*, vol. 11, no. 7, pp. 87–97, December 2010.

[PRE 10b] PRERADOVIC S., KARMAKAR N.C., "UWB chipless tag RFID reader design", *IEEE International Conference on RFID – Technology and Applications*, June 2010.

[RAM 11] RAMOS A., LAZARO A., GIRBAU D. *et al.*, "Time-domain measurement of time-coded UWB chipless RFID tag", *Progress in Electromagnetics*, vol. 116, pp. 313–331, 2011.

[RAM 12a] RAMOS A., GIRBAU D., LAZARO A., "Influence of materials in time-coded chipless RFID tags characterized using a low-cost UWB reader", *42nd European Microwave Conference*, pp. 526–529, Amsterdam, 2012.

[RAM 12b] RAMOS A., LAZARO A., GIRBAU D. *et al.*, "IR-UWB radar system and tag design for time-coded chipless RFID", *6th European Conference on Antennas and Propagation (EUCAP)*, pp. 2491–2494, March 2012.

[RAM 13] RAMOS A., LAZARO A., GIRBAU D. *et al.*, "Time-coded chipless RFID temperature sensor with self-calibration based on a Vivaldi antenna", *IEEE MTT-S International Microwave Symposium Digest (IMS)*, pp. 1–4, 2013.

[RAM 16a] RAMOS A., LAZARO A., GIRBAU D. *et al.*, "Signal processing techniques for chipless UWB RFID thermal threshold detector detection", *IEEE Antennas and Wireless Propagation Letters*, vol. 26, pp. 618–621, 2016.

[RAM 16b] RAMOS A., PERRET E., RANCE O. *et al.*, "Temporal separation detection for chipless depolarizing frequency-coded RFID", *IEEE Transactions on Microwave Theory and Techniques*, vol. 64, no. 7, pp. 2326–2337, July 2016.

[RAM 16c] RAMOS A., LAZARO A., GIRBAU D. *et al.*, *RFID and Wireless Sensors Using Ultra-Wideband Technology*, ISTE Press Ltd, London and Elsevier Ltd, Oxford, 2016.

[RAN 17] RANCE O., PERRET E., SIRAGUSA R. *et al.*, *RCS Synthesis for Chipless RFID*, ISTE Press Ltd, London and Elsevier Ltd, Oxford, 2017.

[REJ 13] REJA M.M., HAMEED Z., MOEZ K. *et al.*, "Compact CMOS IR-UWB transmitter using variable-order Gaussian pulse generator", *Electronics Letters*, vol. 49, no. 16, pp. 1038–1040, 2013.

[REZ 13] REZAIESARLAK R., MANTEGHI M., "Short-time matrix pencil method for chipless RFID detection applications", *IEEE Transaction on Antennas and Propagation*, vol. 61, no. 5, pp. 2801–2806, May 2013.

[REZ 15] REZAIESARLAK R., MANTEGHI M., "Accurate extraction of early-/late-time responses using short-time matrix pencil method for transient analysis of scatterers", *IEEE Transactions on Antennas and Propagation*, vol. 63, pp. 4995–5002, 2015.

[RIZ 16] RIZWAN M., KUTTY A.A., KGWADI M. *et al.*, "Comparative study of inkjet and thermal printing for fabrication of passive UHF RFID tags", *10th European Conference on Antennas and Propagation (EuCAP)*, pp. 1–5, Davos, 2016.

[SAT 12] SATTLEGGER K., DENK U., Navigating your way through the RFID jungle, Texas Instruments white paper, February 2012.

[SCH 05a] SCHWOERER J., MISCOPEIN B., UGUEN B. *et al.*, "A discrete fully logical and low-cost sub-nanosecond UWB pulse generator", *IEEE Annual Conference Wireless and Microwave Technology*, pp. 43–46, 2005.

[SCH 05b] SCHANTZ H.G., *The Art and Science of Ultra-Wideband Antennas*, Artech House Print on Demand, 2005.

[SCH 09] SCHÜßLER M., MANDEL C., MAASCH M. *et al.*, "Phase modulation scheme for chipless RFID- and wireless sensor tags", *Asia Pacific Microwave Conference*, pp. 229–232, Singapore, 2009.

[SCH 14] SCHUSSLER M., KOHLER C., WIENS A. *et al.*, "Screen printed chipless wireless temperature sensor tag based on Barium Strontium Titanate thick film capacitor", *SENSORS*, pp. 2223–2226, November 2014.

[SHE 12] SHEN Y., LAW C.L., "A low-cost UWB-RFID system utilizing compact circularly polarized chipless tags", *IEEE Antennas and Wireless Propagation Letters*, vol. 11, pp. 1382–1385, 2012.

[SMI 17] SMITH P., Little Known Characteristics of Phase Noise, Analog Devices Application Note AN-741, 2017.

[TAM 12] TAMMAM E., YANG L., ALLAM A. *et al.* "Design of a small size UWB antenna with band-rejection characteristics", *IEEE Conference on Electronics, Communications and Computers*, pp. 112–117, 2012.

[TED 12] TEDJINI S., PERRET E., VENA A. *et al.*, "Mastering the electromagnetic signature of chipless RFID system", in KARMAKAR N.C. (ed.), *Chipless and Conventional Radio Frequency Identification: Systems for Ubiquitous Tagging*, IGI Global, 2012.

[TED 13] TEDJINI S., KARMAKAR N., PERRET E. *et al.*, "Hold the chips: chipless technology, an alternative technique for RFID", *IEEE Microwave Magazine*, vol. 14, pp. 56–65, July 2013.

[VEN 11] VENA A., PERRET E., TEDJINI S., "Chipless RFID tag using hybrid coding technique", *IEEE Transactions on Microwave Theory and Techniques*, vol. 59, no. 12, pp. 3356–3364, December 2011.

[VEN 12a] VENA A., PERRET E., TEDJINI S., "Design of compact and auto-compensated single-layer chipless RFID tag", *IEEE Transactions on Microwave Theory and Techniques*, vol. 60, no. 9, pp. 2913–2924, September 2012.

[VEN 12b] VENA A., PERRET E., TEDJINI S., "High capacity chipless RFID tag insensitive to the polarization", *IEEE Transactions on Antennas and Propagation*, vol. 60, pp. 4509–4515, October 2012.

[VEN 13a] VENA A., PERRET E., TEDJINI S., "A depolarizing chipless RFID tag for robust detection and Its FCC compliant UWB reading system", *IEEE Transactions on Microwave Theory and Techniques*, vol. 61, no. 8, pp. 2982–2994, August 2013.

[VEN 13b] VENA A., PERRET E., TEDJINI S. *et al.*, "Design of chipless RFID tags printed on paper by flexography", *IEEE Transactions on Antennas and Propagation*, vol. 61, no. 12, pp. 5868–5877, 2013.

[VEN 14] VENA A., PERRET E., SORLI B. *et. al.*, "Study on the detection reliability of chipless RFID systems", *XXXIth URSI General Assembly and Scientific Symposium (URSI GASS)*, Beijing, 2014.

[VEN 16] VENA A., PERRET E., TEDJINI S., "Design of Chipless RFID Tags", *Chipless RFID based on RF Encoding Particle*, ISTE Press Ltd, London and Elsevier Ltd, Oxford, 2016.

[VIO 05] VIOLINO B., "A summary of RFID standards", *RFID Journal*, January 2005.

[WAN 12] WANG Y., UWB pulse radar for human imaging and doppler detection application, PhD thesis, University of Tennessee, Knoxville, 2012.

[WHY 08] WHYTE G., DARBARI F., MCGREGOR I. *et al.*, "Different feeding geometries for planar elliptical UWB dipoles, and the excitation of leakage current", *38th European Microwave Conference*, pp. 1382–1385, 2008.

[WIE 09] WIESBECK W., ADAMIUK G., STURM C., "Basic properties and design principles of UWB antennas", *Proceedings of the IEEE*, vol. 97, no. 2, pp. 372–385, February 2009.

[XIE 06] XIE H.L., WANG X., WANG A. *et al.*, "An ultra-low-power low-cost Gaussian impulse generator for UWB applications", *IEEE International Conference on Solid-State and Integrated Circuit Technology Proceedings (CSICT)*, pp. 1814–1820, 2006.

[XIL 11a] XILINX, LogiCORE IP RAM-based Shift Register v11.0, available online at: https://www.xilinx.com, 2011.

[XIL 11b] XILINX, LogiCORE IP FIFO Generator v8.2, available online at: https://www.xilinx.com, 2011.

[XIL 15a] XILINX, Spartan-6 FPGA Clocking Resources User Guide UG382 (v1.10), available online at: www.xilinx.com, 2015.

[XIL 15b] XILINX, Spartan-6 FPGA SelectIO Resources User Guide UG381 (v1.7), available online at: www.xilinx.com, 2015.

[XIL 17a] XILINX, LogiCORE IP XPS UART Lite (v1.02.a), available online at: https://www.xilinx.com, 2017.

[XIL 17b] XILINX, XILINX Device Driver API, available online at: https://www.xilinx.com, 2017.

[ZHA 06] ZHANG L., RODRIGUEZ S., TENHUNEN H. *et al.*, "An innovative fully printable RFID technology based on high speed time-domain reflections", *Conference on High Density Microsystem Design and Packaging and Component Failure Analysis*, pp. 166–170, Shanghai, 2006.

# Index

## A, B, C

## D, E, F



## G, H, I

## L, M, N



## O, P, Q



## R, S, T

## U, V, W, X

FSC
www.fsc.org
MIX
Papier | Fördert gute Waldnutzung
FSC® C083411

Kolibri 360 GmbH
Ferdinand-Jühlke-Straße 7
99095 Erfurt, Deutschland
produktsicherheit@kolibri360.de